What on Earth?

Teacher Guide

DEVELOPED BY
EDUCATION DEVELOPMENT CENTER, INC.

KENDALL/HUNT PUBLISHING COMPANY
4050 Westmark Drive P.O. Box 1840 Dubuque, Iowa 52004-1840

This book was prepared with the support of National Science Foundation (NSF) Grant ESI-9255722. However, any opinions, findings, conclusions and/or recommendations herein are those of the author and do not necessarily reflect the view of NSF.

Copyright © 1998 by Education Development Center, Inc. (EDC).

ISBN 0-7872-2215-1

All rights reserved. No part of this book may be reproduced in any form or by any electronic or mechanical means, including information storage and retrieval systems, without permission in writing from the publisher. Notwithstanding the foregoing, teachers may duplicate the reproducible masters in this book without charge for the limited purpose of use in their classrooms.

Printed in the United States of America
10   9   8   7   6   5   4   3   2   1

ACKNOWLEDGEMENTS

What on Earth? module was developed by Education Development Center, Inc. (EDC), with funding from the National Science Foundation.

What on Earth? module was conceived and written by the staff at the Center for Science Education at EDC.

EDC DEVELOPMENT TEAM
Jacqueline S. Miller – Principal Investigator
Judith Opert Sandler – Project Director
Amy R. Pallant – Associate Project Director
Grace S. Taylor – Senior Associate

Kyle O'Hannon – Research Associate
Kristin Metz – Research Associate
Kristina Garrett Ransick – Research Assistant
Kristen Bjork – Research Associate

PRODUCTION
Kerry S. Ouellet – Production Coordinator
Carol Stone – Copy Editor
Lisa Fowler – Designer
Lou Genovese – Illustrator

ACKNOWLEDGEMENTS

The following teachers were involved in the testing of *What on Earth?* module in their classrooms and greatly contributed to its development:

Lucille Andolpho
Mt. Hope High School
Bristol, Rhode Island

Lornie Bullerwell
Dedham High School
Dedham, Massachusetts

Candace Dunlap
Malden High School
Malden, Massachusetts

Peter Geary
Westwood High School
Westwood, Massachusetts

Maureen Havern
Cambridge Rindge and Latin High School
Cambridge, Massachusetts

Randy Johnson
Lynnfield High School
Lynnfield, Massachusetts

Dorothy Quinn
Holliston High School
Holliston, Massachusetts

Frank Scafati
Wellesley High School
Wellesley, Massachusetts

JoAnn Staiti
Westwood High School
Westwood, Massachusetts

TEACHER/SCIENTIST ADVISORY PANEL

James Hamos, Office of Science Education, University of Massachusetts Medical School
Kaaren Janssen, Managing Editor, Current Protocols in Molecular Biology
Elin Kaufman, Harvard Medical School
Sandra Mayrand, Worcester Foundation for Experimental Biology
Donald Bockler, Arlington High School, Arlington, Massachusetts
Richard Howick, Belmont High School, Belmont, Massachusetts
Susan S. Plati, Brookline High School, Brookline, Massachusetts
Grace S. Taylor, Cambridge Rindge and Latin High School, Cambridge, Massachusetts

The staff would like to acknowledge the contributions of the following individuals: Karen Worth, Senior Scientist, Center for Science Education, Education Development Center, Inc., for support and guidance in both the curriculum development process and in new approaches to teaching and learning; Dr. Max Terman, Department of Biology, Tabor College, and Dr. Eric Pallant, Department of Environmental Science, Allegheny College, for reviewing the materials for scientific accuracy; Rene Kruter for the cover photograph of the cichlid; and Jerrie Hildebrand for format design for the pilot version.

TABLE OF CONTENTS

I. INTRODUCTION TO *INSIGHTS IN BIOLOGY*
Philosophy and Goals i
Design of the Curriculum ii

II. THE FRAMEWORKS
Teaching/Learning Framework iii
Science Thinking and Process Skills Framework iv
Assessment Framework iv

III. COMPONENTS OF THE MODULE
Organization of the Teacher Guide vii
Special Features of the Module viii
The Student Manual ix
The Student Notebook ix

IV. TEACHING STRATEGIES
The Classroom as a Community xi
Cooperative Learning Groups xi
Concept-Mapping xii
Models xv
Technology Tools xv
Discussion xvII
Inquiry xix
Critical Thinking xx
Classroom Safety Rules xx

V. MODULE DESCRIPTION
Overview of *What on Earth?* xxiii
Purpose xxiv
Outcomes xxiv
Assumptions of Prior Knowledge and Skills xxiv

	Annotated Table of Contents	xxvi
	Calendar	xxix
	Complete List of Materials	xxxiv

VI. Introductory/Final Questionnaire — xxxv
    Scoring the Questionnaire — xxxvii

VII. Learning Experiences

1	Home Is Where the Habitat Is	1
LTP	*Long-Term Project—Investigate Locally, Think Globally*	11
2	Oh, What a Tangled Web	17
3	Round and Round They Go	37
4	Population Pressures	53
	For Further Study: Be Fruitful and Multiply?	65
5	Variation...Adaptation...Evolution	77
6	The Diversity of Life	93
	For Further Study: Going...Going...Gone!	105
7	Back to Nature	117

VIII. Appendices

A	Materials List–Materials Needed by Learning Experience	125
B	Resource List	129

Introduction

Introduction

INSIGHTS IN BIOLOGY

▶ PHILOSOPHY AND GOALS

Insights in Biology, an introductory biology course for students in the ninth or tenth grade, is designed as an inquiry-based curriculum which focuses on providing all students with the ability to understand fundamental principles in biology and with the skills to apply these understandings in their everyday lives.

In electing to create a new approach to biology education, Education Development Center, Inc. has recognized that society's need for a scientifically literate citizenry has increased dramatically in the last few decades, and that a solid knowledge base in the biological sciences is critical for the health and well-being both of the individual and of our society. In addition, in order to succeed in the technology-based, information-intensive society of today individuals must possess certain process and critical thinking skills that will enable them to seek out the knowledge they need and apply this knowledge to new situations. To prepare for the challenges of this kind of society, students must acquire and practice these skills in their educational experiences. With this perspective in mind, the goals of *Insights in Biology* include:

- providing a core understanding of essential knowledge in biology by exploring specific concepts in depth and in a context relevant to the student;
- presenting each concept in several different contexts to demonstrate the unifying themes in biology and to emphasize the connected nature of biology;
- connecting the understanding of biological principles with applications to day-to-day living, with social, economic, and ethical issues, and with the responsibility for making decisions about one's own health and environment;
- developing the skills to transfer the knowledge of these concepts to new situations and to apply them appropriately;
- developing critical-thinking and problem-solving skills through direct experience in the laboratory, in research investigations, in inquiry-based activities, in role-playing, and in case studies;
- providing support to teachers in developing new instructional strategies and pedagogical

INTRODUCTION

- approaches that are interactive, inquiry-based, hands-on, and that develop science process and thinking skills;
- fostering in each student the willingness and skills required to engage in inquiry—the ability to address a problem, phenomenon, or issue and to reach an understanding, conclusion, or opinion using knowledge and critical-thinking and problem-solving skills; and
- cultivating in students an appreciation for the aesthetic value of biology as well as for its roles in technological advances and in everyday decision-making in the home, community, and workplace.

▶ DESIGN OF THE CURRICULUM

Insights in Biology is designed with the belief that students must have a variety of learning experiences, each affording them opportunities to ask and answer meaningful, thoughtful, and challenging questions about themselves and their world. The design of the curriculum also strives for a flexibility that accommodates different teaching and learning styles while maintaining high standards of conceptual understanding. In order to achieve these goals the curriculum design includes:

- a format consisting of five modules which can constitute a complete core curriculum for introductory biology, or can be used as separate units in combination with other programs or curricula;
- the option to sequence modules in a way that reflects the interests, conceptual development, and needs of the teacher and his or her particular group of students;
- the presentation of fundamental biological concepts in a context of major topics, demonstrating themes of biology and the interrelatedness of these concepts;
- content that addresses the role of the biological sciences in developing new technologies which affect our options and decisions in modern life;
- pedagogical approaches that support the teacher in:
  - responding to student interests and experiences
  - developing instructional strategies
  - helping students make conceptual connections
  - supporting students in the development of inquiry skills
  - developing student driven experiences
  - assessing student understanding
- instructional materials that are inquiry-based, activity-oriented, and that incorporate manipulatives, model building, and quantitative analysis as a bridge between concrete and formal thinking in a cooperative learning environment;
- learning experiences that provide opportunities for students to explore specific concepts in greater depth and to identify interests and ideas, and pursue them in a variety of activities; and
- assessment strategies that inform instruction and measure student concept and skills development.

Frameworks

Frameworks

INTRODUCTION

II. THE FRAMEWORKS

In order to assist in the implementation of an inquiry-based curriculum, *Insights in Biology* is organized around three explicit frameworks: a teaching/learning framework, a science thinking and process skills framework, and an assessment framework.

▶ TEACHING/LEARNING FRAMEWORK

The teaching/learning framework has a dual purpose: first, it supports the conceptual flow of the module; second, it provides a structure for each individual learning experience (lesson). This framework is a guide to the phases of the learning experiences and provides a basis for implementing effective inquiry strategies such as questioning, discussing, creating hypotheses, and working in groups. The phases of a learning experience in this curriculum include: Setting the Context; Experimenting and Investigating; Processing for Meaning; and Applying. One learning experience may include all or only some of the phases.

SETTING THE CONTEXT This phase engages students in the materials and concepts that they will be investigating. It provides teachers with the opportunity to assess students' prior conceptions; it provides students with an interest base and a motivation to explore further; and it connects each learning experience to the storyline of the module and to concepts that were developed in earlier learning experiences. This phase may take the form of a discussion, demonstration, or specific task.

EXPERIMENTING AND INVESTIGATING During this phase, students take on problems or challenges. The activity or investigation forms a basis of common experience upon which students can build new concepts and perceptions. This phase provides students with opportunities to develop new problem-solving and inquiry skills. The activities include laboratory experimentation, model building, and case studies. In some instances students use inquiry skills to design experimental approaches to the curriculum's questions or to those of their own making.

PROCESSING FOR MEANING In this phase, students organize the data, ideas, and experiences obtained in the activity. They use their data to come to conclusions and to process the experience as a means of developing new understandings of concepts. Students hone their communication skills as they explain their results and conclusions and place them in the context of the module's themes.

APPLYING This phase challenges students to apply their understanding to new situations and to extend the concepts and process skills that they have used. Students may be called upon to explain a phenomenon related to a particular learning experience, to develop another approach for examining a concept, or to solve a problem related to what they have been doing, but in a different context. This phase requires students to make connections to other biological concepts, to their lives, and in some instances, to other content areas.

INTRODUCTION

▶ SCIENCE THINKING AND PROCESS SKILLS FRAMEWORK

In addition to an in-depth investigation of science topics and technology, the curriculum provides for the development of higher-level thinking skills and the formation of inquiry-based approaches to learning. Thus, the curriculum helps to create a structure in which students can learn to question, research, verify and organize information, analyze and synthesize conclusions, explore alternatives, and make choices. As students investigate science concepts in the context of inquiry-based exploration and problem solving, higher-level thinking skills are embedded and support the ultimate goals of helping students become critical thinkers and life-long learners.

▶ ASSESSMENT FRAMEWORK

Assessment strategies in *Insights in Biology* are designed for two purposes:

- to assist in informing instruction by providing insight into student progress through each learning experience;
- to provide measures for determining how closely students have come to achieving the objectives of the learning experiences and the outcomes of the module.

Assessment is a broader concept than testing and includes any strategies by which progress is made toward a particular set of goals—in this case, the goals of this curriculum. The primary purpose of assessment in education should be instructional decision-making. Decisions may be short-term or long-term. You might need to decide, for example, how you would change your approach to the next class session. Or, a long-term decision would be whether a student is ready to move on to the work of the next module. To make both kinds of decisions, you must have a sense of where students are presently, and of the nature of the work and demands that lie immediately ahead.

Assessment for instructional decision-making means obtaining the information about where individual students are at this moment with respect to conceptual understanding, thinking skills, scientific processes, and with respect to where we want them to be by the end of this period of instruction.

The following is a brief discussion of the different assessment tools and strategies in the *Insights in Biology* curriculum. These include strategies that provide you with information on how well your students are understanding concepts and developing their thinking, laboratory, and group skills, so that you can make daily adjustments to your teaching. They also enable you to assess the progress of each student throughout the entire module.

INTRODUCTORY QUESTIONNAIRE The Introductory Questionnaire is administered at the beginning of the module. It is designed to help you determine which, if any, of the module's basic concepts and skills students already understand; which concepts and skills they understand partially; and which ones they do not know at all. The Introductory Questionnaire is intended as a written exercise; however, you are encouraged to supplement or even replace it with interviews if you have students with limited English facility or with special needs.

ASSESSMENT STRATEGIES (found in the margin) Assessment strategies can be found in each of the learning experiences. These assessments consist of questions designed to provide ongoing information that will help you determine the progress of your students through each learning experience.

The goals of each learning experience are diverse. They aim at the acquisition of a particular concept, focus on learning new process skills,

INTRODUCTION

require students to work productively in groups, and target the development of attitudes, such as curiosity, thoughtfulness, and interest in science. The assessment strategies found in the margin help you focus on the specific goals of each learning experience and how the students are thinking about scientific concepts, and help you identify any naive explanations they may presently believe.

The focus of the assessment strategies is to assist you in monitoring both individual and class development. These will help you adapt the learning experiences by revising timing or grouping, reinforcing concepts, or changing teaching strategies. These assessment strategies also allow you to build a cumulative picture of students' growth in concepts and skills.

STUDENT NOTEBOOK Each student is being asked to keep a student notebook which will contain all the student's responses to Analysis questions, laboratory reports and experimental designs, student's own questions, and other writings from the module. The student notebook can serve as an ongoing and permanent record of student progress in conceptual understanding and skill building. As such, the notebook can be used to monitor and assess students as they advance through the module.

EMBEDDED ASSESSMENT This type of assessment allows you to determine how well students are building concepts and developing skills in one or more learning experiences. Examples of these may be student analysis questions, laboratory reports, concept maps, questions designed to determine student integration of concepts from several learning experiences, and long-term projects. Scoring rubrics are provided for each embedded assessment.

A general guide to scoring rubrics is shown below. For each embedded assessment examples of appropriate responses and levels of achievement are provided.

Embedded assessments are found in this module in Learning Experiences 3, 4, 6, and in the Long-Term Project.

SCORING GUIDE FOR EVALUATION OF STUDENT RESPONSES

SCORE	DESCRIPTION
Level 4	Student completes all of the important components of the task and communicates effectively. The response demonstrates in-depth understanding of the relevant content and/or procedures. Where appropriate, student chooses more efficient and/or sophisticated procedures and goes beyond the minimum requirement to offer insightful interpretations or extensions (e.g., generalizations, applications, analogies). Student works cooperatively in a group.
Level 3	Student completes most of the important aspects of the task and communicates successfully. The response demonstrates understanding of major concepts even though some less important ideas or details may be overlooked or misunderstood. Student works cooperatively in a group.
Level 2	Student completes some of the important aspects of the task, but not others. Student communicates some important ideas. Gaps in student's conceptual understanding are evident. Student shows some difficulty in working cooperatively.
Level 1	Student completes only a small portion of the required components of the task. The response reveals only fragmented understanding of concepts. Student fails to work cooperatively.
Level 0	Student's response is either totally irrelevant or, if there is some understanding of the task, the response is totally wrong or provides no evidence of appropriate reasoning.

INTRODUCTION

FINAL QUESTIONNAIRE The Final Questionnaire has the same questions as the Introductory Questionnaire. The repetition of the same questions allows you to evaluate the degree of change that has occurred during the module. Its purpose is to help you assess how students have grown in their understanding of key concepts and process skills presented in the module. You may also wish to assess how well students work cooperatively in a group setting. Scoring rubrics are provided.

ASSESSMENT STRATEGIES AND THE ISSUE OF GRADES

It is important to distinguish between the assessment strategies in this module and other kinds of testing, evaluation, and grading that may happen in the school. Tests are traditionally used to evaluate student achievement at the end of a unit or term. They are structured to measure what students know, and a certain cut-off score is considered "passing." The assessments in this module are designed primarily to help you determine what students do not yet understand, or only partially understand, and are intended to enable you to determine your next teaching objectives. These assessments are different than the chapter tests and quizzes associated with textbooks which often provide measures only of what a student has memorized. The assessment strategies in the module are *not* designed to tell you how to evaluate each of your students, nor are they recommended to be the sole provider of all information on which grades are determined.

You are encouraged to develop or use your own method of grading students. While you may glean significant feedback from the identified strategies, there are some aspects of student performance that are not addressed. For example, students are required to answer questions and discuss concepts throughout each learning experience. The evaluation of student work on these pieces will be determined by you, the teacher. You should feel free to produce your own method of evaluation and incorporate it into this program.

There are a number of ways to keep records, including the use of anecdotal records, portfolios of students' work, and checklists of homework and other student materials. We encourage you to include, along with ongoing data collection, a fairly precise record of where students are with regard to concepts, process skills, and group skills at the beginning of the module, at one or more intermediate checkpoints, and at the end. Portfolios, in which students collect and present examples of their best work in required areas, are a particularly effective evaluation tool.

The final questionnaire is more concerned with measuring change and growth than with producing a score. It, therefore, should help you determine whether a student is progressing satisfactorily, one of the pieces of information in your ultimate decisions about a "grade."

Components

Components

INTRODUCTION

III. COMPONENTS OF THE MODULE

The components of the module include the Teacher Guide and the Student Manual. The Teacher Guide contains phases of the Teaching/Learning Framework for each learning experience, the Assessment Strategies, the Introductory and Final Questionnaires, background information, and teacher support materials. The Student Manual has all of the student activity pages and readings.

▶ ORGANIZATION OF THE TEACHER GUIDE

Each learning experience in the *What on Earth?* module follows a similar format:

LEARNING EXPERIENCE INTRODUCTION These pages provide you with an at-a-glance outline of the learning experience, including the following:

Overview: a brief paragraph summarizing what your students will be doing in the learning experience.

Learning Objectives: the science concepts and skills that the learning experience addresses.

Suggested Time: the minimum amount of time you will need to complete the learning experience, based on classroom testing.

Materials Needed: the materials you will need to conduct the learning experience. As appropriate, the list is broken down into three groups: materials for each student, materials for each group of students, and materials for the class as a whole.

Advance Preparation: exactly what you need to prepare beforehand, including special materials, classroom arrangement, and charts.

Technology Tools: a brief description of a particular technology (software, video, CD-ROM) and how it might be used to supplement the curriculum.

Assumptions of Prior Knowledge and Skills: a description of what students should be familiar with before beginning the learning experience.

Teaching Sequence Preview: a summary of "what is happening" in each phase of the Teaching Sequence.

For Further Study: an explanation of the purpose, science concepts, and skills that this additional learning experience addresses, including how this experience is related to previous concepts in the module, the concepts it explores in greater depth, and any new concepts that are introduced. (See Special Features of the module on page viii.)

TEACHING SEQUENCE These pages provide a detailed guide for teaching the four phases of the Teaching/Learning Framework—Setting the Context, Experimenting and Investigating, Processing for Meaning, and Applying. Instructional support is provided by demonstrating the conceptual flow and providing information and questions for implementing the learning experience. The teaching sequence is correlated to the Student Manual.

Number of Sessions: Each learning experience has been divided into sessions to provide an idea of how long each part of the learning experience might take. These are not rigid time frames and should be used only as a guide. Each class may differ in its time requirements.

INTRODUCTION

▶ SPECIAL FEATURES OF THE MODULE

FOR FURTHER STUDY After Learning Experiences 4 and 6, you will find a section called *For Further Study*. Its purpose is to provide additional experiences related to the concepts that students have been exploring in the main sequence of the learning experience. *For Further Study* is designed to enhance and deepen student understanding of concepts already developed in the module, introduce topics that extend student understanding in new areas, and develop student appreciation of new facets of certain biological concepts.

The *For Further Study* experiences include fundamental biological concepts that are connected to the development of the concepts, themes, and storyline of the module, while simultaneously providing you, the teacher, with flexibility about integrating the new concepts and information. At the places where *For Further Study* occurs, you may judge that your students' comprehension of concepts presented in the learning experience is not strong enough to warrant further investigation. In this case you may choose not to take advantage of the *For Further Study*, but to go on to the next learning experience.

In addition to the teacher material, student pages are provided in the Teacher Guide as Blackline Masters for you to photocopy and distribute prior to the class session in which they are to be used. Where the *For Further Study* experiences occur we have listed the sessions in the calendar, as we feel these are important experiences that should be incorporated into your lesson planning.

THE LONG-TERM PROJECT An integral feature of this module is the long-term project introduced after Learning Experience 1. This project is designed to engage students in an ecological or environmental field study in the area in which they live. The project helps students apply concepts from the module to the ecosystem they will be studying and focuses on developing process skills including: observing and describing, collecting data, analyzing, drawing conclusions, and communicating information effectively.

The project may be conducted simultaneously with the learning experiences. It will be most effective if you consider the following details before you introduce the project:

- **Amount of time:** You will need to determine the amount of time students will need to carry out a meaningful research project. In order to determine the time, you may want to consider: site visits, daily data collection, library research, visiting local businesses, and in-class time for students to work on their data analysis and final reports.

- **Ordering materials:** Order materials for student groups as early as is possible.

- **Use and organization of a project laboratory notebook:** A project notebook that each student keeps could be one of the most important parts of the project. In the notebook, students will keep an ongoing record of all their observations, questions, hypotheses, and group discussions. You may wish to spend some time stressing the importance of complete data-taking when the project is introduced.

- **The project field study report:** The project report can be used as an assessment. There is a rubric provided to help you assess students abilities to design a field study, demonstrate field skills, and to write a report. The rubric will need to be adjusted according to the project. When the project is introduced, you may want to discuss the requirements with students and remind students about the importance of doing quality work on the project. Stress that careful attention and detail as the project proceeds will undoubtedly make for greater understanding and a higher quality final report.

CASE STUDIES In Learning Experiences 1, 4 and 7, and in the two *For Further Studies*, you will find bioethical case studies which are an integral, interwoven part of this module. The topics were chosen because of their relationship to the concepts that have been developed in the learning experiences in

INTRODUCTION

which they appear. In these case studies, students are challenged to ask themselves, "What ought I do?" In replying to this question, students analyze the situation involved, examine their own values, and apply their conceptual understandings.

Once students have read the case study, the Analysis and discussion that follow will help them focus on what they have read, form individual opinions, and make their own decisions about the larger issues the cases represent.

Using case studies, students develop critical-thinking and decision-making skills. The following components should be included in a student's response to a bioethical case study:
- identification of the problem
- explanation of the ethical issues involved
- identification of as many solutions as possible
- analysis of current and future consequences of each possible solution
- choice of personal solution ("What ought I do?")
- justification of their solution
- values they have which helped them with their decision-making

In Learning Experience 1—Home is Where the Habitat Is, there are suggestions on how to use the case study as a teaching tool and strategies for engaging students in bioethical decision-making (see the Teaching Strategy in Session Five of the learning experience).

▶ THE STUDENT MANUAL

The Student Manual includes the readings and activities students use in exploring the concepts in biology and in developing inquiry and critical thinking skills. The readings consist of materials from magazines, newspapers, books, and original writings. The activities include laboratory experimentation, simulations, and model building.

Each learning experience includes a section entitled Extending Ideas. One component of Extending Ideas is On the Job, which describes career opportunities that relate to the concepts and materials students have been exploring. This section provides students with information about the kinds of jobs that are available, the level of education needed to pursue this type of job, and the nature of the work involved.

Extending Ideas also provides suggestions for research and activities that can be used to extend conceptual understandings in the learning experience, apply the concepts to new situations, explore related ideas and areas, and to find out about historical or community related events that relate to the biological concepts students have been examining.

▶ THE STUDENT NOTEBOOK

The student notebook is an individual notebook—loose-leaf, spiral, or bound—which each student needs to use for all written assignments in the module *What on Earth?* These assignments will include writing responses to Analysis questions following readings and activities in the Student Manual, writing laboratory reports, and composing essays. The notebook is also the place to keep individual records of charts, tables, and any other notes used during the course of the module. All assignments and records should be labeled according to the learning experience in which they occur.

The student notebook represents an integral part of the student learning cycle. You may wish to discuss with your students the importance of using it for all their written work. A well-maintained notebook will enable students to assess the progress of their own conceptual understandings as they move through the module and to monitor their improvement in skills such as writing and critical thinking. Students should also view their notebooks as a resource enabling them to refer back to their ideas and thinking in previous learning experiences.

Teaching Strategies

Teaching Strategies

INTRODUCTION

IV. TEACHING STRATEGIES

▶ THE CLASSROOM AS A COMMUNITY

Significant influences on learning include the use of a variety of classroom instruction techniques and the climate within a classroom. Positive teacher-student and student-student interactions, instructional and social, contribute to feelings of self esteem and foster a sense of belonging in the class and the school (Wang, 1993). The kinds of camaraderie that exist in a successful sports team—the enthusiasm and drive to succeed—should be the goal of the classroom teacher as well.

DEVELOPING THE CLASSROOM COMMUNITY In a curriculum such as this, which is not textbook-driven but emphasizes the hands-on, minds-on approach to learning, classroom climate plays a critical role. Incorporating activities that promote students' learning about and from each other is a good way to begin the year and is time well spent. Name games that are expanded to include other pertinent information, such as hobbies or sports, are always useful. Another introductory activity is to have each member of the class (including yourself!) bring in and explain three objects that are meaningful to him or her. This type of sharing reveals the student as a person. For this activity it is important to arrange the classroom in a way that everyone shares eye contact. Connecting with the extracurricular activities of students, such as participation in a school play or concert during the year, builds on and maintains a sense of community.

Since the format and methodologies of this curriculum may be unfamiliar to students—in that "correct" answers are derived from experiences, not given in a teacher's book—students will be needing support and encouragement. As students begin to know and trust each other and as cooperative groups form and succeed, you will no longer need to be the sole source of recognition and appreciation of student talents. In stressing the "we" nature of learning, the members of your classroom will become a community of curious and thoughtful investigators.

Wang, Margaret C., et al. "What Helps Students Learn?" Educational Leadership, Dec 1993/Jan 1994, pp. 74–79.

▶ COOPERATIVE LEARNING GROUPS

BUILDING CONFIDENCE Teacher experiences and research show that students in cooperative learning groups gain more confidence in their ability to do science and also learn in a more integrated and conceptual way (Posner and Markstein, 1994). Developing effective cooperative groups takes more structure than telling four students that they form a group. It involves new skills for you and your students. In order to focus students' efforts, you may want to identify a specific task. To facilitate student interactions, help students to define their roles in that task and to understand the importance of positive supportive interactions. Students should also be made aware that each individual is accountable for the group's work. As students become more adept working with groups, they may take on most of these tasks themselves. A goal of this program is to present mechanisms that enable students to take responsibility for their own learn-

INTRODUCTION

ing. Productive cooperative groups encourage and sustain student initiatives. Cooperative learning groups support the kinds of attitudes that were mentioned in The Classroom as a Community section. They reduce student apprehension and isolation.

ORGANIZATION OF GROUPS   Most of your work occurs before the lesson actually begins. It is important to formulate the learning objective, accumulate the necessary resources, and assign students to their roles in heterogeneous groups (such as facilitator, reader, recorder, materials/resource person). While all groups may have the same task, there may be times when each group will have a different task and come together at the end to contribute to the whole discussion.

Activities and laboratory investigations in this module have been designed with cooperative learning groups in mind. We feel that as students learn to listen, discuss, and share tasks at hand, they should acquire more confidence and independence. Within each learning experience are suggestions about group size.

The size of the groups varies according to the learning task required. There is no right way to form cooperative learning groups. In some cases you may want students to choose their own groups. In others you may want to construct the groups to best fit the class. The following are guidelines that may help you divide your class into groups:

- Each group has students whose performance levels range from low to high.
- The average performance level of every group is about equal.
- Groups are balanced appropriately as to ethnicity, race, and gender.

It is suggested that groups stay together for more than one learning experience. In this way, each student could have an opportunity to be in a different role within the group.

ACCOUNTABILITY   Accountability for the finished product may take various forms. You may want one result from each student or one from the group. If you use the latter method, each student must sign the completed report, signifying that he or she knows and understands the contents. As the groups work, your role is that of a monitor—listen for positive student interactions and for questions that students use for clarification and understanding. It is very important that each group have enough time to process how their group functioned and to discuss ways they might improve. As you and your students become more proficient at cooperative learning, you will all find it an important teaching and learning method that benefits students academically and socially.

Posner, Herbert B. and James A. Markstein. "Cooperative Learning in Introductory Cell and Molecular Biology." *Journal of College Science Teaching, February 1994, p. 231.*

▶ CONCEPT-MAPPING

Concept maps are diagrams of the connections between related concepts. The development of concept-mapping skills is an integral part of the *Insights in Biology* curriculum and is used in each of the five modules. We have chosen to emphasize it for several reasons. Concept-mapping is useful to both students and teachers. It helps students organize information, identify relationships between concepts, see gaps in their own understanding of these relationships, and make the connections to relevant experiences outside of class. Student concept maps enable teachers to identify students' prior conceptions and to assess student understanding. Concept-mapping can also help teachers introduce or review concepts visually.

MAKING CONCEPT MAPS   A concept map usually illustrates one main concept, identifies key related concepts, and describes the connections between

INTRODUCTION

these concepts. Concepts (usually nouns) should be circled and the lines drawn between the circles labeled with one or two words (usually verbs or prepositions) that establish the defining connections between them. In the example below, "plant" is the main concept; related concepts include "water," "carbon dioxide," and "solar energy;" connecting arrows indicate that plants use water and carbon dioxide, and absorb solar energy. Each circled concept should be linked to at least one other concept. Cross-linkages may connect concepts in different strands of the map and demonstrate deeper understanding of the interrelationships among them.

Discourage students from constructing maps that are strictly linear (example B below) in favor of maps that include branches and cross linkages (example A below). Many students have problems with 'stringiness' when they start making concept maps. If you find students stringing more than two or three concepts in a row, they probably should consider connecting that second or third concept further up on the map.

Emphasize to your students that no single map is the correct map. But while there may be no "right" maps, there are ones that are incomplete or make incorrect or inaccurate connections. Be careful that your students make maps that say what they mean. They should recheck their map as they construct it. Does each concept connect to each one above and below it with a meaningful linkage on that line shared between them? It is important for students to explain their reasoning.

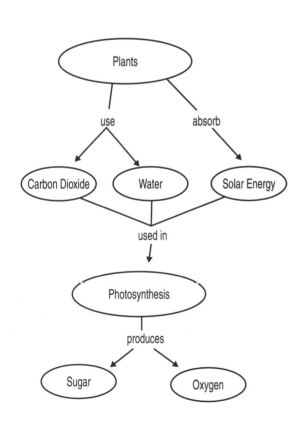

Example A

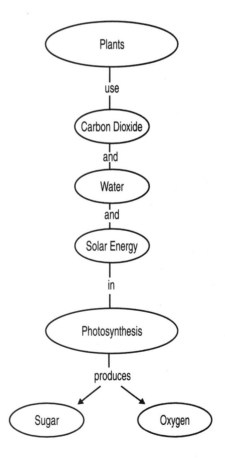

Example B

INTRODUCTION

SUGGESTED DO'S AND DON'TS FOR STUDENTS
- Do not use more than three concepts in a row without branching out.
- Check to make sure the hierarchy of subconcepts makes sense.
- Make sure linking words are appropriate.
- Use no more than three words in a linking phrase.

TEACHING STRATEGIES Teachers introducing concept-mapping for the first time may find it difficult. It is a tool that will be unfamiliar to many, and it may take students some time before they become comfortable with it and proficient enough to fully demonstrate their knowledge of a subject. In the beginning they may need encouragement to stick with it. Your taking it seriously, showing enthusiasm, and insisting that they give it their best effort will make a big difference. Many students and teachers who struggled with it at first have found concept-mapping to be exceptionally valuable once the skill is developed.

INTRODUCING CONCEPT-MAPPING You may wish to introduce concept-mapping by having the class begin with one or more familiar subjects before mapping unfamiliar concepts such as photosynthesis or biomolecules. (Familiar subjects might include: basketball; pizza; the school community; the music industry, or any other subjects familiar to your students.) You may wish to begin by brainstorming words related to the main concept and then have students make connections to the main or other concepts. Another technique for helping students learn how to construct a map is to have them write concepts and subconcepts on self-adhesive paper notes so that they can easily be moved around as students change their minds about how various concepts should be linked to each other.

BUILDING THE SKILL GRADUALLY Some teachers have found it helpful to begin by creating concept maps as a whole class exercise until most students have developed a level of comfort with the process and then having students map concepts in small groups, before finally asking students to produce concept maps on their own. This process might be done over a period of several days, weeks, or months depending on students' comfort and skill level. For a while, as students are developing their skill, you may want to provide them with a list of the key concepts that should be included in their maps, allowing them to concentrate on making the appropriate linkages between these concepts. For example, you may tell them that you expect a map of "ecosystems" to include biotic factors, abiotic factors, community, niche, food chain, producer, consumer, and any other related terms that you expect them to understand.

Eventually students will be able to identify the key concepts themselves. Concept-mapping is included throughout the *Insights in Biology* curriculum; in this module it appears in Learning Experiences 2, 3, 5, and in the *For Further Study* that follows Learning Experience 6. As students work through the modules, their skills in concept-mapping should increase dramatically. Concept-mapping should be used in places other than those suggested by the curriculum. You may wish to use this approach to identify assumptions or prior knowledge students bring to a topic, to develop a small- or large-group consensus about a concept or theme in biology, or to help students assist each other in understanding a concept.

EVALUATING CONCEPT MAPS Concept maps are valuable for informing you about a student's progress with the concepts presented in a reading, in an activity, or in a laboratory investigation. Incomplete or inaccurate connections serve to alert you to what needs further explanation and clarification. Or they may be used to assess student knowledge instead of, or as a part of, laboratory reports, quizzes, or exams, or in class activities.

Since student maps will vary, rather than looking for conformity, you should look for the characteristics of good concept maps: organized

INTRODUCTION

from general to specific, more branched than linear, links and concepts distinguished, logical relationships shown between concepts, and appropriate linking words and logical cross linkages.

▶ MODELS

MODELS AS A LEARNING TOOL The use of models is integral to many of the module's learning experiences. The usefulness of a model depends on how well students are able to imagine how the model represents something they do not understand or can not see in the classroom, and how it is like something they do understand.

EXAMPLES AND USES OF MODELS In some instances, models are physical or structural constructions. A physical model is an actual three-dimensional representation or process that symbolizes or behaves enough like the phenomenon being modeled that some understanding can be gained from it. Your students undoubtedly have built model airplanes, constructed model towns from building sets, or built model castles in the sand. Students should understand that their "playthings" are symbols of real-world items. Adults in general, and scientists in particular, frequently use models to help them make sense of complex ideas. James Watson and Francis Crick, for example, built their first DNA model out of building pieces that were very much like tinker toys.

It is important, while using these physical models, to have students analyze the behavior of the models. Students should be asking: What are the expected limitations of the models? A model cannot be expected ever to represent the full-scale phenomenon with complete accuracy, not even in the limited set of characteristics being studied. It is therefore important to draw students into analyzing the model itself.

In biology, models are often used for understanding biological systems. For example, in this module, students model the complex feeding relationships among a community of organisms. In the activity, each student represents a population of organisms in a specific community, and with string, connects to another student that represents its food source. As each member completes the connections, the class creates a food web. Students then simulate with the model the impact of a toxic substance on the environment.

Models may also involve the use of analogies in order to give an unfamiliar thing meaning by likening it to something familiar. In this module, students observe unfamiliar human-made objects to determine the function of each object from its structure. Students then explore the importance of this structure-to-function principle in the plant and animal world.

EXTRAPOLATION FROM MODELS Too often students come away from a science class thinking that atoms are miniature solar systems, that cells are just compartments like cells in a honeycomb, or that genes are capital and small letters. When models appear in the module, take the time in class to discuss the question of "what is reality" and to look closely at how in each case a model is used as a tool to understand something.

The kinds of models used are dependent on the situation, the kind of concept being understood, the ethics, the timespan, or other difficulties encountered using the real thing. You may wish to have the students explore and analyze the model, or even come up with their own models. In this way, you can assess the success of the model while simultaneously seeing how students understand the concept they are trying to model and the reasons for using the model.

▶ TECHNOLOGY TOOLS

Good technology tools, including software, videotapes, and videodiscs, can introduce or reinforce concepts, tutor students, give feedback, and keep track of progress. Each technology can enhance

INTRODUCTION

learning in a different way. In this way, technology adds another dimension to the teacher's repertoire of strategies. The importance of using technology in science classrooms is not limited to the ability to improve learning. Rather it adds another dimension to instructional strategies, as well as promotes familiarity for most students who will be required to be technology literate for their livelihoods.

COMPUTER USE IN THE SCIENCE CLASSROOM

Technology is not an integral component of the *Insights in Biology* curriculum. A deliberate decision was made to ensure that the use of the curriculum did not depend on the availability of technology. Computer usage by any teacher is a function of his or her computer experience and expertise, availability of hardware and software, and perceived need. An excellent science course can be taught without the use of a computer. However, the careful incorporation of computers into a science course can and does add an important level of enhancement (ERIC Digest, 1991). As students interact with computers in a variety of ways within their science courses their degree of computer awareness and literacy will increase (ERIC Digest, 1991).

The most widely used applications are word processing programs, spreadsheets, and drawing programs. Most recently networks such as the Internet and World Wide Web have made available vast amounts of information. A small but increasing number of science teachers are using computers in selected laboratory activities.

USES TO ENHANCE OVERALL LEARNING Biology activities can use technology tools in their investigations. Technology may provide tools that help collect data, manage information, perform calculations, and that create graphs, drawings, and images. In addition, there are interactive programs that include simulations for building and exploring computer models of phenomena. The simulations in these programs serve as surrogates for phenomena that are too hazardous or time-consuming or could not be provided better through other means in most high school biology classrooms. Well-designed simulations enable students to open the black box of phenomena that are normally inaccessible and see how things work by manipulating variables. However, such innovations should not replace direct experience—activities, field trips, and hands-on activities. Because there are shortcomings to these simulations, combining hands-on and simulated experiences offers the best of both worlds—the concrete experience of laboratory work and the benefits of being able to conduct experiments in a faster mode. Any technological innovation should be used in ways that take advantage of its capabilities, so that it is different from reading a book.

Recommendations for software, videodiscs, and videotapes appropriate to specific learning experiences are provided in the Resource List in Appendix B. Informational guides for using networks such as the Internet and World Wide Web are also provided in the Appendix.

Computer Uses in Secondary Science Education. *ERIC Digest.* ERIC Clearinghouse on Information Resources, Syracuse, NY 1991, EDO-IR-91-1.

THE INTERNET The Internet is a network of thousands of interconnected computers that allows users to access information and files. A wealth of information about many subjects is available through this vast network. The Internet can serve as a powerful tool and a valuable resource for finding the most up-to-date information on many subjects addressed in the *Insights in Biology* curriculum. Students doing projects related to specific learning experiences within the module or wishing to pursue research in the Extending Ideas sections of the module should be encouraged to use the Internet, if access is available. Students can access libraries around the world, have discussions with scientists, read newspaper and magazine articles, exchange ideas or even data with other students around the world, and experience a wide range of perspectives and opinions on scientific topics.

INTRODUCTION

Many on-line service providers are available for accessing information from the Internet. The following services allow you access to many aspects of the Internet:

Newsgroups are places in the Internet where special interest groups can exchange ideas. Messages can be posted in these electronic "bulletin boards" for others to read and post their replies. Large numbers of newsgroups devoted to many subjects are available on the Internet and can be accessed through on-line service providers and through the World Wide Web (see below).

Gopher is a database and communications system that runs on the Internet. Search engines (tools that are used to search the Internet) are available which allow you to do keyword searches of a wide variety of topics.

Telnet is useful for searching the electronic catalogs of libraries. Access to these libraries requires a Telnet program such as NCSA Telnet or Windows Telnet. An educational database (ERIC) can be accessed using the address sklib.usak.ca (password eric). Articles pertaining to education can be searched for by author, title, keyword, and subject.

The **World Wide Web (WWW)** is a very popular graphical interface that allows access to the Internet. Access to the Web is available through a wide variety of web browsers currently available at low or no cost including Netscape Navigator, Mosaic, MacWeb, and the Internet Explorer. The Web allows the transmission of words, pictures, sounds, and movies, a dimension which makes available to users a rich, multi-media resource.

▶ DISCUSSION

LEARNING THROUGH DISCUSSION Classroom discussions form an integral strategy for teaching and learning in the *Insights in Biology* curriculum.

Discussion questions appear throughout the curriculum. They are used as an entry into new learning experiences to set the context; as a way for students to analyze their investigation and reach new understandings and as a tool to determine student understanding of the concept; and as a means of having students apply these understandings to new situations or to determine what you might need to do to ensure student understanding of the concepts. The questions provided in the curriculum are *examples* of the type of questions you may wish to ask as a way of initiating and guiding the discussion. They are, by no means, the only questions, and as facilitator of discussions you should be prepared to guide the discussion in the directions that the students take it, keeping in mind the intent of the discussion as described in the text that surrounds the suggested questions.

DISCUSSION AS AN ESSENTIAL SKILL The ability to conduct dialogue and to exchange ideas and knowledge is an essential life skill as well as being of primary importance in science. In the scientific community, constructive exchange of ideas provides a critical foundation for developing new ideas and innovative approaches to exploration and discovery. Whereas talking arises naturally in most humans, the ability to participate in thoughtful and useful discussions must be learned. Classroom discussion provides students the opportunity to practice this skill and develop an understanding of the importance of discussion in the learning process. In this curriculum classroom discussion facilitates student learning by providing them with the opportunity to:
- clarify their ideas in their own minds;
- articulate their own ideas and make them explicit by explaining ideas;
- relate their ideas to those of other students and to different contexts;
- expand and develop their ideas through dialogue with others;
- determine whether their current ideas contra-

INTRODUCTION

dict what is being learned, and explore why this might be; and

- develop related skills such as listening, participating, examining ideas for logic and evidence, identifying implications and consequences of making statements, using examples and analogies to clarify ideas, and being able to distinguish what is known from what is believed.

DISCUSSIONS TO INFORM INSTRUCTION Classroom discussion also informs instruction by enabling teachers to:

- assess prior student understanding;
- determine the level of student understanding of the concepts;
- assess student ability to apply understandings in different contexts;
- determine reasoning behind student conclusions; and
- inform instruction based on student thinking.

Use student responses to determine whether misconceptions are occurring and to decide whether concepts should be revisited or explorations of concepts continued. Discussion may identify pathways for future investigations and explorations which you or your students may wish to pursue as a group or individually.

Questions may be posed by the students themselves and flow from a prior assignment. Or, you may pose your own questions, or use the questions in the curriculum to determine current knowledge and to stimulate interest in a topic. In every case, the questions should be broad and open-ended—or designed to help "sum up" a learning experience—and allow for a wide range of thinking.

FACILITATING AND CREATING A SAFE ENVIRONMENT
It is critical that you create a safe environment for discussion by ensuring that students listen and consider the viewpoints of others, and that no comments or opinions are cut off, ignored, or unfairly dismissed. Let students know that in order for interactive debate and conversation to take place, it is important that the entire class create an environment that is comfortable for everyone to be able to present their opinions. It may help to open the discussion by asking students to describe what a safe environment would be for them. You may wish to have the class brainstorm a list of ground rules for discussion and then determine on which items on the list the class agrees.

TECHNIQUES TO ENCOURAGE DISCUSSION Both you and your students should practice "wait time." Such purposeful pauses before asking for a response increase both class participation and the quality of thinking (Rowe, 1979). Getting everyone to speak early-on encourages more participation and thoughtful analysis later.

It is always preferable that small-group discussions be followed by a whole-group discussion so that all the ideas may be presented and incorporated. You should redirect questions to students to encourage and foster student-student interchanges. At intervals, students should be asked to link and then summarize the key points or hypotheses. Whenever applicable, specific references from the reading or experiment should be given. Always keep in mind that discussions may diverge depending on student questions, answers, elaborations, and clarifications.

An effective discussion, like any other successful teaching methodology, requires planning. Discussion should not be chaotic or a free-for-all but should be structured enough to take student thought toward clarity, consistency, accuracy, and relevance. Your role is as facilitator rather than leader. As facilitator it is your task to pose thought-provoking questions; to break large questions into smaller, more manageable ones; to help students clarify their thoughts by rephrasing or asking different questions; to help keep the discussion focused; to encourage students to explain their comments to each other: and to recognize the need for new information or knowledge when that is called for.

Rowe, Mary Budd (1979). "Wait, wait, wait . . ." *School Science and Mathematics.* 78: 207.

INTRODUCTION

▶ INQUIRY

COMPONENTS OF INQUIRY   Science education has often stressed "the scientific method" or inquiry as a pedagogical approach that models the way scientists solve problems. However, inquiry learning can be thought of in a much broader framework as a strategy or set of strategies for living; for asking good questions, for designing ways of answering these questions, for solving problems, and for making decisions based on information and concepts relating to the question or problem. These strategies require certain skills that can be developed in the science classroom but can be applied throughout life. The components and skills that define an inquiry-based approach to learning include the abilities to:

- ask an appropriate question about an inherently interesting phenomenon;
- define what aspect of that question you want to find out about;
- determine what you already know and what else you might need to know;
- conjecture or hypothesize about the solution to the question or problem;
- gather knowledge, data, experience, and information that might help answer the question or solve the problem;
- analyze the information and determine whether it is accurate and sufficient for your purpose;
- draw a conclusion based on the information;
- apply the result to a broader question to see if the conclusion can be generalized or used in other decision making situations; and
- develop new questions based on these conclusions.

DEVELOPING INQUIRY SKILLS   The challenge to teachers in using an inquiry approach is in identifying contexts in which inquiry can be meaningfully placed. Students need to experience each component of the process. Students should learn to analyze evidence and data and to discuss their experiences with peers in order to develop their thinking and understanding. Using an inquiry-based approach to learning, one can enter the process at any stage. For example, the question may be provided but the learner must gather information or design and perform the experiment, collect and process the data, and come to a conclusion. Or the learner may be provided with data and asked to analyze it, draw a conclusion from it, and develop new questions based on this conclusion. Students can develop the specific aspects of inquiry and then combine all the components in a single investigation. This approach provides students with greater understandings of the concepts with which they are working and the skills they need to apply these understandings to new situations.

   In the science classroom, inquiry is a way of teaching as well as learning. Students should be encouraged to ask questions, and you as the teacher should be able to say, "I don't know, let's find out together. What information do we know, what information do we need . . ." thereby modeling the inquiry process.

INQUIRY INVOLVES DEPTH RATHER THAN BREADTH
The emphasis in learning through inquiry is depth rather than breadth; the inquiry approach provides opportunities for students to delve more thoughtfully into a topic, to deepen their understandings of fundamental concepts, to make the connections among concepts and principles. Using an inquiry-based approach in the biology classroom recognizes that coverage of all the facts and concepts usually addressed in traditional biology curriculum cannot occur. However, this approach develops the critical-thinking skills students will need for lifelong inquiring and learning.

INTRODUCTION

▶ CRITICAL THINKING

DEVELOPING REASONING AND PROBLEM-SOLVING SKILLS The explosion of information in biology in recent years has forced teachers to look at their discipline in new ways. It is impossible to teach all that is currently known. The resulting shift in curriculum planning and pedagogical approach includes a focus on major, fundamental concepts and instruction that develops students' abilities to use and apply these concepts in an analytical and new way.

Critical thinking has been defined as "reasonable, reflective thinking focused on deciding what to believe and do" (Ennis, 1989). Your role as teacher is to help your students develop reasoning skills and problem-solving strategies. This is a complex task and involves deliberate and explicit efforts.

STUDENTS AS ACTIVE LEARNERS Critical-thinking skills are embedded within the module's learning experiences. An explicit goal is for students to become active learners. As they gain a base of knowledge they are asked to pose questions, devise hypotheses, identify lab controls, collect and interpret data, and draw valid conclusions. It is important for students to understand how these conclusions were drawn and to think through further ramifications and interpretations that may arise from these conclusions.

A suggested entry point is to devise activities that sharpen students' observation skills. An example is to place a variety of objects on a tray, have a group of students observe the objects for thirty seconds, and then have the students list as many as they recall. To follow up, ask the students to group the objects into four or five logical categories and include their rationale. To extend the activity further, ask which specific real-life situations demand careful observation and categorization of objects for information. Apply these skills in many diverse situations and contexts to make them integral parts of your students' thinking processes.

Another introductory activity is to ask students to critique advertisements. The following questions should prompt observations: What are the conclusions? What data is given to support these conclusions? What are the assumptions? What are the inferences? What do you predict the results of this ad have been? What experiment would you devise to test your prediction? Was this experiment attempted? How do you know? Is your experiment valid? How do you know? Concepts and facts gained in this manner form an interrelated biological knowledge base. They are used to explain new observations, information, and conclusions that are then added to form a more coherent and complete network of thought. Teaching strategies mentioned in other sections, such as concept-mapping, discussion, and inquiry, also reinforce and support student critical thinking.

Ennis, R. H. "Critical Thinking and Subject Specificity: Clarification and Needed Research." Educational Researcher, *1989, 18: 4–10.*

▶ CLASSROOM SAFETY RULES

GUIDELINES FOR CLASSROOM SAFETY Safety is of primary concern in conducting investigations and activities in the biology classroom. Each student should be well informed of safe practices during laboratory investigations, and you should have a classroom safety program that outlines safety procedures, potential hazards, and appropriate behavior in a laboratory setting. Careful preparation and attention to recommended precautions and procedures are important features of preventing laboratory accidents. Students and teachers should also be well acquainted with the steps to be taken in the event of an accident. The following is a list of basic safety considerations and steps that should form the basis of the safety program in your classroom.

INTRODUCTION

1. Obtain and be familiar with a copy of the federal, state, and local regulations that relate to school safety as well as a copy of your school district's policies and procedures.

2. Prior to every investigation, check to be sure that all recommended safety equipment is available. During the investigation check that the safety equipment is being used properly by the students. Keep all aisles clear.

3. Before beginning an investigation, review any specific safety features of the investigation. Areas requiring specific safety considerations are framed by a box marked "Caution" or "Safety" and are highlighted in the margin within the Student Manual.

4. All equipment not in use should be stored in a separate, locked location when you are not present. All chemical hazards, such as acids or bases, should be kept in rubber buckets away from areas of student activity.

5. Prior to investigations, call students' attention to any special safety equipment in the room such as the eye wash, safety shower, and fire blanket. Be sure that students are instructed in the use of this equipment.

6. Inform students that they are not permitted to work in the laboratory without an instructor present. Inform students never to taste any of the materials of the investigation. No eating, drinking, or smoking is permitted in the laboratory area.

7. Plan sufficient time to clean up after each activity; have a separate, labeled disposal container for glass and sharp objects and separate, labeled containers for each chemical waste reagent. All microbial growth should be disposed of by autoclaving or treating with household chlorine bleach before discarding. Have students check with you before disposing of any materials in a sink.

8. All accidents or injuries, no matter how small, should be reported immediately to the appropriate authorities.

CLASSROOM SAFETY RULES The following is a list of safety procedures to post in the laboratory area:

1. Do not mix unknown chemicals just to see what happens.

2. Do not touch nor taste any chemical unless specifically told to do so by your teacher.

3. Do not smoke, drink, or eat in the laboratory area. Behave in a serious manner.

4. Always wash your hands immediately after using chemicals.

5. Report all accidents, no matter how small, to the teacher.

6. Do not touch your face, mouth, ears, or eyes while working with plants, animals, or chemicals.

7. Do not taste or smell any unknown substance; when you are asked to smell a substance; do so only by gently waving a hand over it to draw the scent toward your nose. Never pipette by mouth.

8. Always follow the safety procedures described in the procedure.

9. When cutting with razor blades or scalpels, use caution; always cut on a stable surface and cut away from your body.

10. Always clean up your work space and yourself after each learning experience investigation.

11. If you have long hair, tie it back before lighting a burner.

SPECIAL CAUTIONS

Chemicals—A hazardous chemical is one that has the potential to cause injury unless precautionary measures are taken. These chemicals include poisons, flammables, reactives, and corrosives. When dealing with hazardous chemicals the following measures should be considered:

1. Chemicals should be stored away from student areas, preferably in a locked storage area.

2. If spills should occur, check the Material Safety Data Sheet (MSDS) that comes with the chemical for specific instructions. Wear non-

INTRODUCTION

porous gloves during the cleaning procedure. In general, solid spills can be swept up and placed in a suitable disposal container. The area should then be washed with soap and water. In the case of a liquid spill, check the pH with pH paper and neutralize with small amounts of acid or base. Use absorbent material to soak up the spill. Discard appropriately and wash the area with soap and water.

3. Certain liquids may be poured down the drain when diluted. Check with your local regulations and the MSDS.
4. Flammable substances are solids, liquids, or gases that will burn readily. Be sure to keep all sources of ignition away from flammable substances. Demonstrate to students the location and use of the fire extinguisher, fire blanket, and safety shower. Be sure that students are familiar with emergency measures in the event that a fire occurs.
5. Impress on students the importance of wearing safety goggles and gloves when handling hazardous chemicals. Instruct them in the use of the eye wash and the fire extinguisher.
6. Inform students that when they transfer hazardous chemicals they must be cautious, be careful not to trip and to avoid drips. Instruct students that acid is always added to water and to clean up any drips or spills.

Live Organisms—When using live organisms in the classroom, the following precautions should be taken:

1. All living organisms used in the biology classroom should be treated with respect and care. No organism should be mistreated in the course of an activity.
2. When an investigation using living organisms is completed, careful consideration must be taken as to what to do with the organisms. Consult local experts before releasing the organisms. No living organism should be released into the environment unless it is native to the area.
3. Actively growing microbial cultures should be disposed of by autoclaving or steam sterilizing at 121°C, 15 psi for 15 minutes. Alternatively, add enough full-strength household chlorine bleach or 70% isopropyl alcohol to the agar dish or growing container. Allow at least 8 hours of contact time before disposing of the culture.

First Aid—Prior knowledge of first aid procedures can minimize health problems in the case of a medical emergency. These procedures include:

1. When new chemicals arrive, consult the MSDS for any special instructions regarding spills, splashes or ingestion. Save all MSDS instructions and use them for orally cautioning students before they begin working with the chemicals.
2. For splashes in the eyes, flush the eyes with copious flowing water for at least 15 minutes. Consult a physician.
3. For skin contact, wash with flowing water for at least 15 minutes. Contact a physician if redness, blisters, or irritation develop.
4. Any clothing that becomes contaminated should be removed and the skin beneath washed. Clothes should be laundered before the next wearing.
5. In the case of inhalation of irritating substances, remove the individual to fresh air. Consult a physician if difficult or painful breathing occurs.
6. If any chemicals are ingested, spit the material out and wash out the mouth with copious amounts of water for at least 15 minutes. Contact a physician immediately.

Module Description

Module Description

INTRODUCTION

▼ MODULE DESCRIPTION

▷ OVERVIEW OF *WHAT ON EARTH?*

What can be learned by studying the interactions of all organisms with the environment? The relationships are extraordinarily complicated. How did the enormous range of organisms interacting in all of the diverse environments evolve? This module explores important ecological concepts, such as the interrelationships of the physical and biological factors that form ecosystems, the principles of population growth and the limiting factors that control population growth, and biodiversity on Earth. These concepts serve as a starting place for students to evaluate specific environmental problems, their causes, and their possible solutions.

On the surface, ecology (the environmental science) appears simple and common sensical. However, each ecological concept is linked to principles in biology and chemistry, as well as to other disciplines, making these apparently simple ideas more complex. In this module, activities, readings, and discussions build on the connections and interrelationships in order to develop fully the concepts. Things that may appear simple when first examined, often reveal many layers of complexities when explored more deeply.

Students begin the module by exploring ecosystems and the organisms that inhabit them. They explore biotic and abiotic factors in a specific habitat, determine the interactions among the organisms and the components needed for life to exist, and then set up a model of an ecosystem. As an application to their analysis of ecosystems, students are then engaged in a case study intended to have them think about how human activity is altering the environment and how changing one component of an ecosystem results in change and compensation throughout the ecosystem.

In a long-term project, which is introduced after Learning Experience 1, students carry out an ecological or environmental investigation in a local area where they collect data and relate their findings to concepts they are learning in the module.

Students then study feeding relationships of populations of organisms by exploring food chains and food webs. In addition, students explores how energy and materials are obtained, processed, and moved or cycled among components of an ecosystem.

The module continues by having students study the underlying factors that cause natural populations of organisms in an ecosystem to increase or decrease. Students explore exponential growth or J-curves, a "boom and crash" curve, and a sustaining or S-curve. Students investigate the interrelationships among populations within an environment by simulating predator-prey relationships. Finally students apply their understanding to a case study on the interactions among several populations—mice, deer ticks, gypsy moths, and acorns—in a northern forest in order to predict the potential hazards for Lyme disease in the forest.

Throughout the module, students have been introduced to the diversity of life and are building an understanding of the complex interactions among a great variety of organisms. They then examine how such a variety of organisms has evolved over time. They study how natural selection and speciation have combined to cause the diversity of life on Earth, and how extinctions create niches for new species. Students simulate how variations are the source of natural selection, explore speciation, construct a definition of species, and examine how climatic events and evolutionary forces have shaped the living world.

In the final learning experience, students return to analyzing how understanding ecological principles have allowed ecosystems damaged by human actions to be restored as in the case study in Learning Experience 1, students research and evaluate the solutions to environmental problems. Specifically, students look at the plans to restore the Florida Everglades, and the biological, social, political, and economic issues that surround the restoration of this area, and decide what ought to be done.

▷ PURPOSE

How are organisms interrelated with each other and with the environment? How did such complex ecosystems and the diverse organisms in them come to be? What is the future of life on Earth? *What on Earth?* examines the basic concepts that underlie the interactions of organisms, including humans, with the resources in the environment. It introduces the following concepts: The components of ecosystems, how nutrients and energy move through ecosystems, the factors that influence population growth, and how evolutionary forces have caused the diversity of ecosystems and of the species within them.

▷ OUTCOMES

- Students observe and identify the interactions among organisms and their environments and determine that organisms are dependent on one another and their environment for food, other resources, and energy.
- Students create and interpret graphs in order to describe the dynamics of population growth and identify how interspecific population interactions and resource use affects population size.
- Students determine how the processes of natural selection, speciation, and extinction are related to species diversity.
- Students investigate the impact of the relationship between human action and the environment and evaluate these actions in relation to the economic, social, and political ramifications.

▷ ASSUMPTIONS OF PRIOR KNOWLEDGE AND SKILLS

- Students are familiar with the characteristics of life, the metabolic processes of organisms, and the resources needed for life.
- Students understand that plants make their own food, using the light energy from the sun and carbon dioxide from the air, and water from the soil in the process of photosynthesis.
- Students are aware that food provides organisms with energy as well as nutrients.
- Students recognize that microorganisms are part of the environment.
- Students are able to use percentages in calculations.
- Students are aware that calories are a measure of the amount of energy released when food is oxidized.
- Students are familiar with the concept of a closed system.
- Students understand the use of a control in an experiment.

INTRODUCTION

- Students are familiar with chemical indicators.
- Students are aware that inorganic elements in the environment provide organisms with the materials they need to maintain life.
- Students are familiar with creating and interpreting data charts and graphs.
- Students understand that a gene is a segment of a DNA molecule that codes for a specific protein. This molecular transfer of information determines the characteristics of organisms.
- Students are familiar with concept-mapping.
- Students are familiar with using a library and/or the World Wide Web for conducting research.

INTRODUCTION

▶ ANNOTATED TABLE OF CONTENTS

LEARNING EXPERIENCES

1. Home Is Where the Habitat Is
Earth is a world populated by millions of different species of organisms. It is a dynamic world in which change is the norm and relationships among organisms and their environments are constantly shifting and adjusting. This module focuses on both the interactions among organisms and their dependence on the abiotic environment to develop student understanding of the dynamic responses of living things to change in their environment.

This learning experience poses the following questions: What is an ecosystem? How do biotic and abiotic factors interact within an ecosystem? Students explore these concepts as they examine a specific habitat, determine the interactions among the organisms there, identify the components needed for life to exist, and set up model ecosystems. In addition, a case study about the reintroduction of wolves into Yellowstone Park encourages students to think holistically and to examine how changing one component of an ecosystem results in change and compensation throughout the entire ecosystem.

Long-Term Project—Investigate Locally, Think Globally
Students will best understand environmental issues by immersing themselves in a specific, local process or problem. In the long-term project, students carry out an investigative study in a local area, collect data, and relate their findings to concepts they are learning.

2. Oh, What a Tangled Web
In what ways do living things interact in a community? In this learning experience, students will look in greater detail into the role of a population of organisms in a community and focus on the relationships it has with other populations. In this learning experience, students use a grocery store, its workers, and its products as an analogy for organisms linked by food relationships within a biological community. They then identify examples of food chains and examine a marsh ecosystem food web.

3. Round and Round They Go
Sustaining life in an ecosystem requires a steady supply of the resources that organisms need to carry out fundamental life processes such as growth, movement, reproduction, photosynthesis, and respiration. In this learning experience, students explore several essential resources, including nutrients, energy, carbon, oxygen, hydrogen, nitrogen, and water. Students are introduced to the concept of trophic levels and calculate the diminishing amount of energy available in each succeeding trophic level. They then set up an experiment to trace the pathway of carbon and, by inference, oxygen between Elodea and a snail in a closed system. Finally, they examine biogeochemical cycles and determine the importance of the carbon–oxygen, nitrogen, and water cycles to the maintenance of life on Earth.

4. Population Pressures
What underlying factors cause natural populations of organisms in an ecosystem to increase or decrease? What patterns can be discerned and what is the significance of these patterns? In this learning experience, students examine the factors that influence the growth or reduction of populations in an ecosystem over time. Students begin by exploring several different growth patterns: the exponential or J-shaped curve, in which populations with ample resources increase through rapid reproduction; the "boom and crash" curve, in which an isolated population increases rapidly and then decreases quickly; and the sustaining or S-curve in which the population grows rapidly for a time and then

levels off. Students carry out a simulation of predator–prey relationships and analyze a case study about the interactions among several populations—mice, deer ticks, gypsy moths, and acorns—in a northern forest.

For Further Study—Be Fruitful and Multiply?
If human populations continue to grow at their present rate, how will that growth affect environmental quality, ecosystem, Earth, and human life? In this learning experience, students create a graph showing human population levels throughout history. After examining this graph, students discuss factors that contribute to the characteristics of the growth curve and discuss what this information might mean when thinking about Earth's carrying capacity.

5. Variation... Adaptation... Evolution
In this module, students have been exploring the interactions among various organisms and their environments. Whether organisms succeed in these interactions depends on their structural and functional adaptations; adaptations that have evolved over time and led to the variety of life on Earth. But how did such a variety of organisms come about and why did multitudes of other organisms become extinct? The process of evolution, principally by natural selection, is the underlying tenet of biological science: it provides an explanation both for the unity and diversity of life and for the adaptation of organisms to their environment. In this learning experience, students explore how the structures of organisms are correlated to their functions; learn how adaptations, as a form of evolutionary change, arise and are related to the environment; and determine that a group of organisms capable of interbreeding constitute a species.

6. The Diversity of Life
How do new species of organisms arise? By what mechanisms have species changed in the 3.8 billion years that life has existed on Earth? In this learning experience, students explore how new species arise (speciation) and relate this to Darwin's theory of natural selection and finally to biodiversity. Students then examine maps that illustrate patterns of species diversity (richness) in North and Central America and determine the reasons the patterns occur as they do.

For Further Study—Going...Going...Gone!
Does it really matter whether species are endangered and eventually lost? In this learning experience, students examine data that suggests species are disappearing at a rate greater than would be predicted from normal evolutionary causes and identify possible reasons for this accelerated extinction rate. After reading a case study on the Delhi Sands fly, students decide whether species extinction really matters.

7. Back to Nature
By some accounts, humans have disturbed or altered more than half of the world's terrestrial ecosystems through pollution, environmental degradation, resource depletion, and population pressures. What does this mean for humanity? In this learning experience, students will explore in greater depth the intersection of the actions of humans and other species. Can the damage humans have done be repaired? Students will explore how advances in ecological understandings may allow ecosystems damaged by human actions to be repaired. Specifically, students will look at the plans to restore the Florida Everglades and at the issues that surround the restoration of this area.

INTRODUCTION

▷ CALENDAR OF LEARNING EXPERIENCES

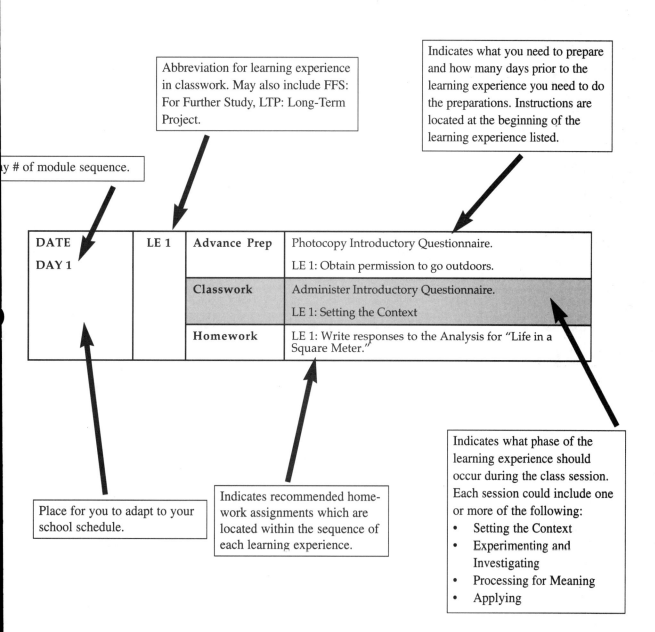

INTRODUCTION

	DATE DAY 1	LE 1	Advance Prep	Photocopy Introductory Questionnaire. LE 1: Obtain permission to go outdoors.
			Classwork	Administer Introductory Questionnaire. LE 1: Setting the Context
			Homework	LE 1: Write responses to the Analysis for "Life in a Square Meter."
	DATE DAY 2	LE 1	Advance Prep	
			Classwork	LE 1: Setting the Context
			Homework	LE 1: Read "The Factors of Life" and write responses to the Analysis.
	DATE DAY 3	LE 1	Advance Prep	
			Classwork	LE 1: Experimenting and Investigating
			Homework	LE 1: Gather materials and organisms for "Life in a Jar."
	DATE DAY 4	LE 1	Advance Prep	
			Classwork	LE 1: Experimenting and Investigating
			Homework	LE 1: Read the case study "Wolves" and write responses to the Analysis.
	DATE DAY 5	LE 1	Advance Prep	
			Classwork	LE 1: Processing for Meaning
			Homework	
	DATE DAY 6	LTP	Advance Prep	
			Classwork	LTP: Explain the Long-Term Project.
			Homework	
	DATE DAY 7	LE 2	Advance Prep	LE 2: Collect core soil samples. Assemble materials for "Mystery Soil." Collect rotting logs.
			Classwork	LE 2: Setting the Context
			Homework	LE 2: Write Responses to the "Mystery Soil" Analysis.
	DATE DAY 8	LE 2	Advance Prep	LE 3: Order materials.
			Classwork	LE 2: Experimenting and Investigating
			Homework	LE 2: Observe organism for 10 minutes and write essay.
	DATE DAY 9	LE 2	Advance Prep	LE 2: Cut yarn or rope. Photocopy one set of biota cards and create a cardboard "sun" for "What's for Lunch?" LE 3: Order Materials.
			Classwork	LE 2: Processing for Meaning
			Homework	LE 2: Write responses to "What's for Lunch" Analysis. Read "Managing Mosquitoes" and write responses to the Analysis.

INTRODUCTION

	DATE DAY 10	LE 3	Advance Prep	
			Classwork	LE 3: Setting the Context
			Homework	LE 3: Create an "ecosystem" concept map.
	DATE DAY 11	LE 3	Advance Prep	LE 3: Prepare a 0.1% bromothymol blue aqueous solution.
			Classwork	LE 3: Experimenting and Investigating
			Homework	
	DATE DAY 12	LE 3	Advance Prep	LE 3: Set up test tubes.
			Classwork	LE 3: Experimenting and Investigating and Processing for Meaning
			Homework	LE 3: Write responses to "An Infinite Loop" Analysis.
	DATE DAY 13	LE 3	Advance Prep	LE 3: Prepare transparencies and make copies of the cycles.
			Classwork	LE 3: Applying
			Homework	LE 1: Write a laboratory report for "Life in a Jar."
	DATE DAY 14	LE 3	Advance Prep	
			Classwork	LE 1: Applying ("Life in a Jar")
			Homework	
	DATE DAY 15	LE 4	Advance Prep	LE 4: Read the *For Further Study* "Be Fruitful and Multiply?"
			Classwork	LE 4: Setting the Context
			Homework	LE 4: Complete the graphs and write responses to the "Unsupervised" Analysis.
	DATE DAY 16	LE 4	Advance Prep	LE 4: Cut hare and lynx cards.
			Classwork	LE 4: Experimenting and Investigating
			Homework	LE 4: Write responses to "The Lynx and the Hare" Analysis. LE 3: If "It's Elemental" is completed, write responses to the Analysis.
	DATE DAY 17	LE 4	Advance Prep	LE 4: Photocopy non-text illustration of "Acorn Rollercoaster."
			Classwork	LE 4: Processing for Meaning and Applying
			Homework	LE 4: Read "Ticks and Moths. . . " and start to design the poster.
	DATE DAY 18	FFS	Advance Prep	FFS: Photocopy Student Pages.
			Classwork	FFS: Setting the Context
			Homework	FFS: Read "Putting the Bite on Planet Earth" and write responses to the Analysis.

INTRODUCTION

	DATE DAY 19	FFS	Advance Prep	
			Classwork	FFS: Processing for Meaning
			Homework	
	DATE DAY 20	LE 5	Advance Prep	LE 5: Locate mystery objects.
			Classwork	LE 5: Setting the Context
			Homework	LE 5: Read "Evolutionary Theory: Past and Present" and write responses to the Analysis.
	DATE DAY 21	LE 5	Advance Prep	LE 5: Photocopy "Movement Directions."
			Classwork	LE 5: Experimenting and Investigating
			Homework	LE 5: Write responses to the "Going with the Flow" Analysis.
	DATE DAY 22	LE 5	Advance Prep	
			Classwork	LE 5: Processing for Meaning
			Homework	LE 5: Read "The Fundamental Unit" and write responses to the Analysis.
	DATE DAY 23	LE 5	Advance Prep	
			Classwork	LE 5: Applying
			Homework	LE 5: Read "Cichlids Past and Present." Inform students that the posters for "Ticks and Moths…." are due in the next session.
	DATE DAY 24	LE 4	Advance Prep	
			Classwork	LE 4: Applying (present posters for "Ticks and Moths….")
			Homework	LTP: Inform students that results of project need to be finalized.
	DATE DAY 25	LE 6	Advance Prep	LE 6: Prepare transparencies of evolution concept maps.
			Classwork	LE 6: Setting the Context
			Homework	LE 6: Read "New Species" and write responses to the Analysis.
	DATE DAY 26	LE 6	Advance Prep	LE 6: Photocopy the three "Species Richness" maps.
			Classwork	LE 6: Experimenting and Investigating
			Homework	LE 6: Write Martian essay.
	DATE DAY 27	LE 6	Advance Prep	
			Classwork	LE 6: Processing for Meaning
			Homework	LE 6: Write comprehensive essay on evolution and speciation.

INTRODUCTION

DATE DAY 28	FFS	Advance Prep	FFS: Photocopy Student Pages.
		Classwork	FFS: Setting the Context and Processing for Meaning
		Homework	FFS: Read case study "Save the *Fly*? Are you Kidding?" and write an essay.
DATE DAY 29	FFS	Advance Prep	
		Classwork	FFS: Applying
		Homework	
DATE DAY 30	LE 7	Advance Prep	LE 7: Read the Learning Experience. Alert personnel in media and computer centers. Locate large map of Florida Everglades.
		Classwork	LE 7: Setting the Context and Experimenting and Investigating
		Homework	LE 7: Complete Task step 1 of "Anything We Want."
DATE DAY 31	LE 7	Advance Prep	
		Classwork	LE 7: Experimenting and Investigating
		Homework	LE 7: Begin research.
DATE DAY 32	LE 7	Advance Prep	LE 7: Invite colleagues and other students to the presentations at the final session.
		Classwork	LE 7: Experimenting and Investigating
		Homework	LE 7: Continue research and develop presentation.
DATE DAY 33	LE 7	Advance Prep	
		Classwork	LE 7: Experimenting and Investigating
		Homework	LE 7: Continue research and develop presentation.
DATE DAY 34	LE 7	Advance Prep	
		Classwork	LE 7: Experimenting and Investigating
		Homework	LE 7: Continue research and develop presentation.
DATE DAY 35	LE 7	Advance Prep	
		Classwork	LE 7: Processing for Meaning
		Homework	LE 7: Complete step 5 of the Task and write a comprehensive paper.
DATE DAY 36	LTP	Advance Prep	
		Classwork	LTP: Processing for Meaning (NOTE: This session should occur at the completion of the project.)
		Homework	
DATE DAY 37		Advance Prep	Photocopy Final Questionnaire.
		Classwork	LE 7: Administer Final Questionnaire.
		Homework	

INTRODUCTION

▶ COMPLETE LIST OF MATERIALS

(Long-term project materials are listed in Appendix A.)

Amounts are for a class of 32 students, in groups of 2, 3, or 4 when applicable.

GLASSWARE AND GENERAL LABORATORY EQUIPMENT
- 32 pairs of safety goggles
- 8 thermometers
- 16 hand lenses
- 16 dissecting needles (or coffee stirrers)
- 16 magnifiying glasses
- 16 petri dishes
- 8 wax marking pencils
- 8 single-edge razor blades or scalpels
- 16 tweezers
- dissecting microscope
- 32 culture tubes or test tubes with tight-fitting covers or stoppers
- 8 test tube racks
- 8 eyedroppers
- 2 laboratory stands (optional)

CHEMICALS AND BIOLOGICAL CULTURES
- silicone sealant
- soil, water, plants, compost, fruit flies, spiders, snails and/or other small organisms
- 1 or more soil test kits (optional)
- 8 rotting logs (if available)
- 8 zipper plastic bags each containing a sample of soil from one of three or four different locations (see Advance Preparation)
- 1–2 L distilled water (or "aged" tap water)
- 8–16 mL 0.1% bromothymol blue solution (aqueous)
- 32 small water (pond) snails
- 16 sprigs of Elodea
- salt (optional)

GENERAL MATERIALS (INCLUDING GROCERY ITEMS)
- 32 large nails
- 8 awls (or nails)
- 8 scissors
- 8 darning needles or large safety pins
- 8 hammers, mallets, or fist-sized stones
- 8 trowels, spades, 5-cm diameter piece of pipe, or small bulb planters
- 8 large bags or boxes to carry materials
- 16 paper bags (lunch size)
- 1 plastic bag (optional)
- 2 or 3 large shoebox tops
- clear waterproof tape
- cellophane tape
- masking tape
- field guides (plants, insects, other invertebrates)
- reference material including atlases, encyclopedias, and geography books
- 17 biota cards
- 1 cardboard "sun" (see Advance Preparation)
- 800 hare cards (5 cm x 5 cm) (See Advance Preparation)
- 200 lynx cards (10 cm x 10 cm) (See Advance Preparation)
- 10 sets of "Species Richness" maps
- 16 sheets of newspaper
- 32 sheets of unlined white paper (11 x 14 inches, legal size)
- 150 sheets of graph paper
- label or index card
- 8 transparency sheets
- 1 ball of yarn or string
- wick (cotton rope or other absorbent material)
- 16 felt-tip markers
- 32 or more assorted colored pencils
- 8 meter sticks or tape measures
- 8 rulers or tape measures
- 1 spoon (optional)
- 1 knife (optional)
- 2 bowls (optional)
- 18 or more 2-L bottles of the same brand with caps
- 1 1-L plastic bottle with cap (optional)
- 16 "mystery" objects (such as: potato ricer, strawberry huller, olive pitter, dandelion digger, hose nozzle, nutcracker, hardware item, etc.)
- ice cubes (optional)
- 1 apple (optional)
- access to a fluorescent lamp

Introductory/Final Questionnaire

Introductory/Final Questionnaire

VI. Introductory and Final Questionnaire

Student Name _____ Teacher Name _____

Class Session _____

WHAT ON EARTH?

1. People often say, "We should get rid of all the mosquitoes. They are nothing but pests!" Explain, using good scientific reasoning, why this statement is not true.

2. a. A population is a group of individuals of one species living in a given area. What factors may cause one population to grow indefinitely, another to stabilize, and another to die out.

   b. Describe what might happen in each of these scenarios over time.

3. a. Explain what is meant by "biodiversity."

   b. Explain the forces of nature that cause or result in the appearance of species on Earth?

   c. Why is the maintenance of biodiversity important?

4. Earth may be considered one large ecosystem.
   a. Explain what is meant by an "ecosystem."

   b. Choose an ecosystem. Describe the living and nonliving factors within this ecosystem and explain, using appropriate terms and concepts, how they interrelate and interact.

SCORING THE QUESTIONNAIRE

The following are the rubrics for scoring student responses to the questionnaire.

1. People often say, "We should get rid of all the mosquitoes. They are nothing but pests!" Explain, using good scientific reasoning, why this statement is not true.

SCORING GUIDE FOR EVALUATION OF STUDENT RESPONSES

SCORE	DESCRIPTION
Level 4	Student response is clear and detailed and includes all the components listed below. Student fully describes both the role of mosquitoes in their ecosystem and the effects of their removal on other organisms (and the ecosystem) after its removal.
Level 3	Student response is clear and detailed and includes most of the components listed. Student description of the role of mosquitoes in the ecosystem and the effects of their removal are not complete.
Level 2	Student response is less clear and detailed and includes some of the components listed. Student describes either the role of mosquitoes or the effect of their removal.
Level 1	Student shows little understanding of the role of mosquitoes in the ecosystem or the effects of their removal.
Level 0	Student response is irrelevant or nonexistent.

COMPONENTS OF A LEVEL 4 RESPONSE INCLUDE:

Mosquitoes are important in that they:
- play a role in the interrelationships within its ecosystem (fill a niche)
- are part of the interdependencies in food chains

The removal of mosquitoes causes:
- disruption to those organisms who depend on them for food
- an adverse impact throughout the trophic levels
- possible population changes among its consumers thereby altering the ecosystem

INTRODUCTION

2. **a.** A population is a group of individuals of one species living in a given area. What factors may cause one population to grow indefinitely, another to stabilize, and another to die out.
    b. Describe what might happen in each of these scenarios over time.

SCORING GUIDE FOR EVALUATION OF STUDENT RESPONSES

SCORE	DESCRIPTION
Level 4	Student response is clear and detailed and contains all the components listed below Student shows a full understanding of the factors that foster population growth and those that limit it. Student explains the results of each type of growth over time.
Level 3	Student response is clear and detailed and contains most of the components listed. Student relates the scenarios to time frames accurately.
Level 2	Student response is less clear and detailed and contains some of the components listed. Student does not explain the result over time of each scenario.
Level 1	Student shows little understanding of the three kinds of population growth nor what would result over time.
Level 0	Student response is irrelevant or nonexistent.

COMPONENTS OF A LEVEL 4 RESPONSE INCLUDE:

a. Populations grow indefinitely when there are:
   - more births than deaths resulting in a J-curve
   - large amounts of resources including food and space
   - no or few predators
   - few other checks and balances on the system

   Populations stabilize when:
   - births equal deaths resulting in an S-curve
   - resources are adequate to sustain a limited population
   - the carrying capacity is reached
   - predator and prey offset each other

   Populations die out when:
   - deaths exceed births resulting in a "crash"
   - resources are severely limited
   - wastes build up poisoning the environment
   - predators exceed prey
   - overcrowding produces stress

b.
   - Populations that grow "indefinitely" will ultimately exhaust resources in the environment
   - Populations that stabilize reach their carrying capacity and maintain equilibrium
   - Populations that die out open new niches for new populations to fill

INTRODUCTION

3. **a.** Explain what is meant by "biodiversity."
   b. Explain the forces of nature that cause or result in the appearance of species on Earth?
   c. Why is the maintenance of biodiversity important?

SCORING GUIDE FOR EVALUATION OF STUDENT RESPONSES

SCORE	DESCRIPTION
Level 4	Student response is clear and detailed and contains all the components listed below. Student shows full understanding of biodiversity, the causative natural forces, and both the importance and implications of maintaining biodiversity.
Level 3	Student response is clear and detailed and contains most of the components listed. Student shows an understanding of the influence of natural selection and speciation on biodiversity.
Level 2	Student response is less clear and detailed and contains some of the components listed. Student shows little understanding of the influence of natural selection and speciation on biodiversity.
Level 1	Student shows little understanding of biodiversity and its importance. Natural selection and speciation are not mentioned.
Level 0	Student response is irrelevant or nonexistent.

COMPONENTS OF A LEVEL 4 RESPONSE INCLUDE:

a. Biodiversity refers to:
- the variety of species in a given area (and, by extension, on Earth)

b. The forces of nature that cause biodiversity include:
- variations in organisms over time (their gene pool)
- successful adaptations to environmental changes
- natural selection and other evolutionary forces
- environmental (or climatic) changes that lead to geographic and reproductive isolation and eventually speciation
- extinctions that open up niches for new species

c. Biodiversity is important to:
- maintain interdependencies among organisms
- maintain biogeochemical cycling
- have a varied gene pool on which on selective forces act
- preserve the earth as an ecosystem
- preserve useful products that some organisms produce
- offer pleasure and joy in environmental vistas and the beauty and sound of living things

INTRODUCTION

4. Earth may be considered one large ecosystem.
   a. Explain what is meant by an "ecosystem."
   b. Choose an ecosystem. Describe the living and nonliving factors within this ecosystem and explain, using appropriate terms and concepts, how they interrelate and interact.

SCORING GUIDE FOR EVALUATION OF STUDENT RESPONSES

SCORE	DESCRIPTION
Level 4	Student response is clear and detailed and contains all of the components listed listed below. Student is able to explain and describe an ecosystem and discuss how the living and nonliving factors interrelate and interact accurately.
Level 3	Student response is clear and detailed and includes most of the components listed. Student is able to explain and describe an ecosystem and discuss how the living and nonliving factors interrelate and interact accurately.
Level 2	Student response is less clear and detailed and contains some of the components listed. Student explanation and description of an ecosystem and the interactions among living and nonliving factors is flawed.
Level 1	Student response is not clear and not detailed and contains serious flaws.
Level 0	Student response is irrelevant or nonexistent.

COMPONENTS OF A LEVEL 4 RESPONSE INCLUDE:

a. An ecosystem is composed of all the organisms in a community and the nonliving (abiotic) factors with which they interact in this environment.

b. Interrelationships and interactions include:
   - Predator/prey relationships such as:
     – consumers that feed on autotrophs (herbivores) or other consumers (carnivores)
     – decomposers that feed on dead organisms releasing nutrients and inorganic substances for other organisms
   - The cycling of substances through ecosystems maintaining life on Earth, such as:
     – flow of energy from the sun to the producers and the ultimate loss of energy through trophic levels
     – oxygen
     – nitrogen
     – carbon (both as an element and as the gas CO_2)
     – water (as a liquid and in its gaseous form)
   - Other concepts such as:
     – food chains forming food webs
     – habitat and niche with an ecosystem
     – biodiversity helps to maintain an ecosystem

Learning Experiences

Learning Experiences

Home Is Where the Habitat Is

OVERVIEW The earth is a world populated by millions of different species of organisms. It is a dynamic world in which change is the norm and relationships among organisms and their environments are constantly shifting and adjusting. The science of ecology focuses on the relationships among living things and their relationships to the environment, thus linking the concepts of ecology to other sciences such as biology, chemistry, physics, and geology. An understanding of the complex interactions among the structures and functions of the natural world is necessary for interpreting the impact of changes humans have made in the environment. Since Earth's formation approximately 4.5 billion years ago, a myriad of complex interrelations among its changing inhabitants and their habitats have slowly evolved. In addition, natural events, such as falling asteroids and climatic fluctuations, have led to vast changes in the kinds of microorganisms, plants, and animals that inhabit Earth. However, following such events, new organisms and interrelationships have evolved and a new relationship among the living and the nonliving factors has been established. This module focuses on both the interactions among organisms and their dependence on the abiotic environment to develop student understanding of the dynamic responses of living things to change in their environment.

This learning experience poses the following questions: What is an ecosystem? How do biotic and abiotic factors interact within an ecosystem? Students explore these concepts as they examine a specific habitat, determine the interactions among the organisms there, identify the components needed for life to exist, and set up model ecosystems. These ecosystems will be used throughout the module for illustrating various ecological principles.

Finally, a case study about the reintroduction of wolves into Yellowstone Park encourages students to think holistically and to examine how changing one component of an ecosystem results in change and compensation throughout the entire ecosystem.

LEARNING OBJECTIVES

▷ *Students examine a square-meter plot and identify biotic and abiotic factors in this habitat.*

▷ *Students identify an ecosystem as a community of organisms interacting with one another and with the environment.*

▷ *Students design small ecosystems and introduce the organisms and resources necessary for sustaining life within these ecosystems.*

▶ SUGGESTED TIME

- 6 class sessions (45–50 minute periods)

▶ MATERIALS NEEDED

For each group of four students:
- 1 large bag or box for carrying materials
- 1 meter stick or tape measure
- 4 large nails or golf tees
- 1 hammer or mallet or fist-sized stone
- 5 m of string
- 1 thermometer
- 1 trowel, spade, 5-cm pipe or small bulb planter
- 1 sheet of graph paper
- 1 sheet of newspaper
- 2 hand lenses or magnifying glasses
- 1 petri dish
- 1 soil test kit (optional)
- 2 or more clear plastic 2-L bottles of the same brand, with caps
- 1 wax marking pencil
- 1 razor blade or scalpel
- 1 scissors
- 1 awl or nail
- clear waterproof tape
- 1 darning needle or large safety pin
- silicone sealant
- soil, water, plants, compost, fruit flies, spiders, snails and/or other small organisms
- wick (cotton rope or other absorbent material)

For the class:
- 2 or 3 large shoebox tops

▶ ADVANCE PREPARATION

1. If necessary, obtain permission for your class to go outdoors.
2. If possible, suggest that students wear outdoor clothes for the activity "Life in a Square Meter."

▶ ASSUMPTION OF PRIOR KNOWLEDGE AND SKILLS

- Students are familiar with the characteristics of life, the metabolic processes of organisms, and the resources needed for life.

Notes

See page xx in the Introduction section of this Teacher Guide for **Classroom Safety Rules.**

▶ TEACHING SEQUENCE PREVIEW

SETTING THE CONTEXT
- Students examine a square-meter plot in the school yard and identify the biotic and abiotic factors in this ecosystem.

EXPERIMENTING AND INVESTIGATING
- Students set up an ecocolumn as a model for an ecosystem.

PROCESSING FOR MEANING
- Students discuss the case study "Wolves," identify the issues, determine what choice they would make on the reintroduction of wolves to Yellowstone, and identify values they hold that influenced their decision.

APPLYING
- At the conclusion of the "Life in a Jar" investigation (a few weeks after the ecocolumns are set up), students discuss the successes and failures of the ecocolumns.

TEACHING SEQUENCE

Setting the Context

Session One

Begin class by having students read the Prologue in the Student Manual. In this Setting the Context, students explore their own ideas about the meaning of the terms "habitat" and "ecosystem." Although students may be familiar with these terms and even use them, their understanding may be superficial.

> **Prologue:**
> Student Manual page 1.

DISCUSSION QUESTIONS This discussion should give you an opportunity to determine student preconceptions and their current level of understanding. To facilitate this discussion you may wish to ask the following:

- What is a habitat? What are some examples?
- What is an ecosystem? What are some examples?
- What are some of the features and/or components of an ecosystem?
- Why is Earth itself considered an ecosystem?

> See page iii in the Introduction section of this Teacher Guide for background information on the Teaching/Learning Framework and suggestions for implementing this curriculum.

Write student responses on the board or on chart paper. Explain to students that they will be returning to the concept of ecosystem and adding to this definition.

SCIENCE BACKGROUND

Ecosystems are, by nature, dynamic, not static. Populations of organisms and the physical environment are in constant adjustment. Evolutionary forces are always exerting pressures on populations. Changes in rainfall and temperature also determine the rise and fall of populations. True balance, therefore, is the exception.

For more information on the current thinking about the concept of balance in ecosystems, you may want to read "The Application of Ecological Principles in Establishing an Environmental Ethic" by Charles J. Bicak, *The American Biology Teacher*, Vol. 59, No. 4, April 1997, pp. 200–206.

Session Two

> "Life in a Square Meter":
> Student Manual pages 2–4.

Divide the class into groups of four for the activity "Life in a Square Meter." Inform students that they are going outdoors and are to choose a specific area in the school yard where they will observe what organisms live and visit there. Have each group read the Introduction to "Life in a Square Meter" and the materials list and collect its materials before you accompany the class to the school yard. (If there are no grassy areas, any section of weedy soil and/or sidewalk cracks will be suitable.) A neighborhood park or golf course is another alternative. Remind students not to step in any plot chosen by a group. Have each group follow the Procedure in the Student Manual.

TEACHING STRATEGY

Each group should choose an area somewhat close to, but not next to, another group. Recommend to students that they choose an area with different types of vegetation, if possible, such as a tree or shrub, weeds, and grass. All groups must be within the sound of your voice and/or your whistle.

▶ HOMEWORK

Have students write responses to the Analysis following "Life in a Square Meter." Then have them read "The Factors of Life," which continues to develop the concept of ecosystems, and write responses to the Analysis that follows the reading.

> "The Factors of Life": Student Manual pages 4–5.

Session Three

DISCUSSION QUESTIONS Begin this session by having students discuss the homework. Students should be able to articulate more clearly their understanding of the term "ecosystem." As you discuss the analysis, refer to the lists and definition you recorded at the beginning of Session One. To facilitate the discussion, you may wish to ask the following questions (modified from the Student Manual):

- How do the plants and animals in your plot relate to one another?
- How do nonliving factors affect the plot and the organisms in it?
- Is your square-meter plot considered an ecosystem or part of an ecosystem? Explain your response. *(Students should respond that it is part of an ecosystem because their boundary is an artificial one. The ecological processes act on a landscape scale—small plots are only samples of larger communities.)*
- What other information might you need to understand the connections among the organisms and their environment better?

> See page xvii in the Introduction section of this Teacher Guide for suggestions on holding Discussions.

DISCUSSION QUESTIONS Continue by discussing the resources needed for life (modified from the Analysis that follows "The Factors of Life" in the Student Manual).

- Explain three different ways in which organisms in an ecosystem are dependent upon the abiotic factors in the environment.
- What resources do organisms need to stay alive?
- How do autotrophs, heterotrophs, and decomposers interact to maintain an ecosystem? Give an example of each.

Learning Experience 1   Home Is Where the Habitat Is

- Based on your understanding of ecosystems, do you consider one tree an ecosystem? Why or why not? *(Student responses may vary, but a tree may be considered an ecosystem. It offers food and shelter to many types of organisms that interact within its bark and branches.)*

MODULE CONNECTION

If your students have completed the *Insights in Biology* module *The Matter of Life*, you may wish to review the characteristics of life and the resources required by organisms to sustain life.

Experimenting and Investigating

In the investigation "Life in a Jar," students will design and create bottle habitats. The construction of these ecocolumns will require students to apply their understanding about the resources needed to maintain life and the interrelationships among the biotic and abiotic factors.

Begin by having students read the Introduction and the Procedure for "Life in a Jar." Have students work in groups of four and follow steps 1–4 of the Procedure. For homework, students will need to collect the materials to be included in their ecocolumn as well as the bottles for creating the habitats. As you circulate around the room, listen to how students are thinking about the design of their ecocolumn. You may wish to ask questions such as:

- What type of habitat(s) are you designing?
- What abiotic factors are you including?
- What biotic factors are you including?
- What might you need to add to your ecocolumn over time?

You may wish to encourage students to think about the type of interactions that will occur in order to ensure that life in the ecocolumn will be maintained over time. The following teaching strategy details things to watch for and suggests a variety of possible habitats.

THINGS TO WATCH FOR

The setup illustrated in the Student Manual (Figure 1.3) shows a terrestrial ecosystem above an aquatic one, but students do not need to design their ecocolumns in the same way. Rather, each group's column should be original and reflect the ideas of the group. Encourage students to think about the organisms in their own environment and to be creative about setting up the habitats. Soils and water collected from the natural environment will likely contain algae, phytoplankton, seeds and insect larvae, which will cre-

"Life in a Jar":
Student Manual pages 6–8.

ate a rich ecosystem; store-bought soil and tap water will include far fewer organisms. Remind students that columns could be aquatic or terrestrial habitats or a combination of habitats. Student groups may want to take a profile or core sample—for example, from a pond or grassland—to create their habitats rather than collecting individual organisms and materials to build a habitat. If they choose to create their own, suggestions for possible organisms to be included in the bottles follow. This is a limited list. Anything that fits and is not dangerous may be included.

Plants: mosses, ferns, grasses, small ivies, small flowering plants, spider plants, herbs, seeds, cacti, carnivorous plants, aquatic plants
Animals: ants, beetles, worms, isopods, fruit flies, millipedes, frogs, praying mantis, small fish, clams, crawfish

If students include predators such as spiders or carnivorous plants remind them to maintain a food supply for the organisms.

Plan space to house all the bottles, taking into account the light and temperature needs for the sets of organisms. Remind students to water plants in the ecosystem very lightly, as an accumulation of water will leave the soil soggy and plants may rot.

Have students in each group determine what materials they need to collect and who will bring which of these materials to the next class session.

 HOMEWORK

Have students gather the materials and organisms to set up the ecocolumn.

Session Four

Allow a full session for each group to complete the Procedure and the Analysis for "Life in a Jar." Have students place the completed ecocolumns in a stable location that you designate. Remind students that they must observe these ecosystems on a regular basis and record their observations in their notebooks. They should also monitor the abiotic variables—heat, cold, sun—and adjust the bottle location if necessary. They will be assessed on the thoroughness of their observations and the ways in which they connect their observations to the concepts in the module.

> See page xi in the Introduction section of this Teacher Guide for suggestions on setting up **Cooperative Learning Groups.**

TEACHING STRATEGY

This investigation is designed to last several weeks. If an ecocolumn fails to maintain life, encourage the group to determine what factors were missing from their ecosystem or out of balance for that ecosystem. If time remains, have them set up another ecocolumn using their experience and new knowledge. If not, have students analyze why life was not maintained and how they would set up the ecocolumn differently another time. Their analysis should include their understanding of the interactions among the biotic and abiotic factors in an ecosystem.

Allow 10 minutes at the end of class for students to discuss the Analysis with their group partners and write responses in their notebooks.

▶ HOMEWORK

Have students read the case study "Wolves" and write responses to the Analysis. This case study describes the ecological and economic issues surrounding the reintroduction of wolves into their native habitat of Yellowstone National Park.

> "Wolves": Student Manual pages 9–11.

Processing for Meaning
Session Five

Begin the session by having students engage in a discussion of the "Life in a Jar" Analysis. The purpose of the questions and the discussion is for students to consider what factors are needed for the continuation of life in an ecosystem and for them to make modifications in their ecocolumns, if necessary.

This case study encourages students to think about science-related issues that may be controversial and value-laden. The articles and case studies throughout this module will provide students with opportunities to identify their own values, to communicate them to others, to listen openly to people with conflicting views, and to use both rational thinking and emotional awareness in the discussion of controversial issues.

> See page viii in the Introduction section of this Teacher Guide for more information on using **Case Studies**.

TEACHING STRATEGY

One of the main goals of education is to provide students with the skills to become thoughtful, learned citizens capable of making rational decisions based on a solid knowledge base and an understanding of the possible consequences resulting from each decision. Empowering students in personal decision-making has not normally been included in science education which has tended to focus on encouraging students to ask questions and pursue answers in the laboratory. However, there are similarities between scientific research and bioethical decision-making. Both start with a probing question. In each of the case studies in this module, the students are faced with a "real life" situation with many alternatives for solutions; they are asked to place themselves in that situation, weigh the risks and benefits of each solution both for themselves and for society, and answer the question "What ought I do?" (Note that the word "ought" signifies a more moral and active imperative than "should" or "might," and students ought to be encouraged to use it.)

At first, students may show discomfort in discussing their personal views of ecological issues, environmental dilemmas, and the role of technology. You may wish first to pose a dilemma closer to them, such as a friend who shoplifted or a peer who cheated on a test. Students may prefer to start their thinking in groups, in which alternative perspectives and different aspects of the issue will be discussed, before individually stating their own decisions. It is very important for students to realize that they are being asked to explore many possible solutions, and that decision-making is usually not "either-or" but is more complex. As students describe the reasoning behind their decisions, their own personal values—which have, perhaps unknowingly, influenced their decision—such as friendship or honesty, will emerge. These one- or two-word values may be placed on the board and a class value list assembled. Students are not to accept or reject other students' values, but the list may illuminate values the students have not previously considered, and which might be used in their decision-making in other case studies. As students identify and discuss the local and global consequences of each decision made, some students may choose to change their own decisions, and other values will emerge.

Complex issues are not resolved with neat "right "or "wrong" solutions. As students explore more case studies their conceptual understandings and decision-making skills will result in thinking thoughtfully about how ecological principles and environmental concerns impact their lives now and in the future.

The focus of this case study has students consider the ramifications of first removing the wolves and then reintroducing the wolves into the ecosystem. Students also look at the implications of human needs in this case.

DISCUSSION QUESTIONS Have students begin by discussing the wolves' place in the ecosystem. Continue the discussion with questions such as the following (modified from the Student Manual):

- What happened in the Yellowstone ecosystem after the wolves disappeared?
- What do you think will happen in the years to come?
- At which time do you think the ecosystem was or is most stable? Explain your response.
- Who are the stakeholders in this controversy? How will they be affected?
- What are the issues?
- What further information would you need in order to decide whether to introduce the wolves into Wyoming and Idaho?

Have volunteers state how they would decide whether or not to reintroduce the wolves, and what values influenced their decisions regarding whether or not wolves should be added to the Yellowstone ecosystem. Encourage students to discuss the reasoning behind these choices and further information they need in order to make an informed choice.

ASSESSMENT

▶ Do students realize that the ecosystem in Yellowstone adjusted to the absence of the wolves and will need to readjust to their presence?

▶ Are students beginning to recognize what it means for an ecosystem to be self-sustaining?

Applying

Session Six (to take place several weeks after Session Four)

In this session, students will discuss their conclusions based on the successes and failures of the ecocolumns, their analyses and their conclusions. This should occur after students have had enough time to maintain and observe the ecocolumns. By the end of Learning Experience 3, students have learned the ecological principles necessary in order to write a laboratory report with thoughtful conclusions.

DISCUSSION QUESTIONS Gather the students together to discuss what they have learned from their experimental design and observations. To facilitate the discussion you may wish to ask questions such as:

- What range of days did the ecocolumns remain viable?
- What factors influenced the maintenance of life?
- Describe food chains in your ecocolumn.
- Based on your experience, if you were to design an ecocolumn again, how would you do it differently?
- What principles from the module would you use in this redesign?

TEACHING STRATEGY

Ecocolumns that continue to support life should be maintained and observed as long as possible. You may also want to allow students to adjust variables on these ecocolumns in order to determine the effects on the organisms and the ecosystem. Experiments may include adding substances that might affect terrestrial and aquatic systems such as fertilizers, or pollutants (salts, pesticides, acids) or changing the physical factors, such as light, sound, etc. This will enable students to pose new questions and design new explorations. These investigations will deepen their understanding of inquiry-based science and their understanding of ecological and environmental principles. Caution students that chemicals in certain doses may kill organisms, thus destroying their ecosystem.

Investigate Locally, Think Globally

OVERVIEW In this module, students will be introduced to the broad principles of ecology, the study of the interrelationships between organisms and their environment. While specific examples are used, the general principles apply to global environmental issues. Students will best understand environmental issues by immersing themselves in a specific local process or problem. In the following long-term project, students carry out an investigative study in a local area, collect data, and relate their findings to concepts they are learning.

▶ SUGGESTED TIME

- 1 class session (45–50 minute period) to discuss the long-term project
- 1 class session for student groups to share project ideas with the class
- additional class time as needed

▶ RATIONALE FOR A LONG-TERM PROJECT

If education is to encourage students to become involved, literate citizens, it is important that students not only learn ideas and information presented to them, but also learn to investigate ideas on their own and to think critically. This is particularly true when the subject is ecology. Students may read in newspapers and magazines and hear on television that the earth and her creatures are in danger, whether from destruction of the ozone layer, pollution, or overpopulation. They may not be able to evaluate the threats and may feel helpless in the face of these reports. However, students who are involved in an ecological (environmental) study in the area where they live become familiar with the biotic and abiotic factors in the chosen ecosystem. They also have a context within which to evaluate environmental issues.

This long-term project requires students to develop technical skills in addition to being able to apply ecological concepts to their research. Students may be using test chemicals and equipment, doing quantitative

LEARNING OBJECTIVES

▷ *Students choose an environmental project and devise an experimental plan.*

▷ *Students collect data and write a field study report in which they relate what they have observed to the concepts in the module.*

See page viii in the Introduction section of this Teacher Guide for information regarding the **Long-Term Project.**

and qualitative analyses, observing and recording data, and developing good scientific practices. Curiosity and knowledge may lead some students to civic action such as writing letters, attending local environmental meetings, and changing their own habits.

Most importantly, the project allows students to experience the fun and excitement of scientific investigation. By being involved in an ongoing project, students will be actively engaged in problem-solving, not only of the experimental question at hand, but also of the day-to-day issues that arise in any scientific investigation. By conducting a long-term project, students discover that doing science involves active participation, sharing ideas, questioning, critical thinking, and physical work. Some students may even find it fun!

▶ SUGGESTIONS FOR MATERIALS
(varying with the project[s])

- soil test kits including:
  - nitrogen, phosphorous, and potassium test kits
  - acidity and alkalinity test kits
  - microbe study kit
  - microorganisms test kit
- pH meter
- water test kits including:
  - algae pollution kit
  - bacteria pollution kit
  - coliform pollution kit
  - quantitative analysis kit
  - dissolved oxygen water test kit
  - acid rain survey kit
- water filter kit
- water hardness kit
- seawater analysis kit
- limnology water test kit
- lead testing kit
- cameras and film
- sketch pads
- tape recorder
- hand lenses
- jars and bottles
- field guides
- rain gauge
- thermometers

▶ SUGGESTED PROJECTS

The long-term project should either be developed from students' varied interests, experiences, and ideas that they would like to investigate or should take advantage of the school's surrounding areas—such as a stream, meadow, or forest that is suitable for exploration and analysis. You may also wish to contact your local governmental and environmental organizations (watershed, conservation, fish and wildlife, etc.) to ask about ongoing projects and concerns in your area in which students can participate. The following list provides some other examples of ecological and environmental projects that may be of interest to you and your students.

> See page xix in the Introduction section of this Teacher Guide for information on **Inquiry** in the science classroom.

TEACHING STRATEGY

It may be advisable for all students to investigate one area, such as a local stream, but to do individual or group work on different aspects of it. You may wish to have groups do their observations and tests at different locales along the stream. Whatever you and your student group choose, there are safety precautions specific for each locale and for outdoor work in general. You may wish to do the following:
- Obtain permission from school authorities.
- Obtain a release form from parent or guardian.
- Explain the rules, especially that no student should wander off alone. (Set up a "buddy" system.)
- Ask students about allergies, particularly bee stings.
- Choose areas that are safe.
- Carry first-aid and safety equipment.
- Tell a school official where you are going and when you expect to return.

If students are able to return to the same area yearly, data over the years may be compared leading to a rich longitudinal study.

- *Stream Ecology*
    - Assay water quality (chemicals, pH, dissolved oxygen[DO]).
    - Survey biological components (E. coli, benthic invertebrates, macroinvertebrates [caddis fly stone fly, midge]).
    - Determine biological oxygen demand (BOD).
- *Meadow*
    - Measure flora density.
    - Collect temperature and precipitation data.
    - Observe root system relationship to environment.
    - Test soil quality.
    - Survey fauna present.
    - Investigate mutualism (in fungi).
- *Woodland*
    - Identify flora.
    - Carry out an increment bore (age, weather dendrology).
    - Collect temperature and precipitation data.
    - Identify ground cover and organisms present.
- *Mountain*
    - Determine forest ecosystem (canopy).
    - Identify trees at various altitudes.
    - Collect light, temperature, and soil data.
    - Examine ecological succession.
- *The Living Machine* is a project that treats sewage to reach advanced wastewater treatment standards. This may be accomplished in the classroom laboratory using aquaria or large jars with aerators, siphons, and organisms such as plants and invertebrates.

Notes

The concept behind the design involves a circulation of wastewater (cafeteria remains) through the different balanced ecosystem tanks to yield drinking water. There are 20 Living Machines around the world using this technology (carried on by Dr. John Todd, a Canadian biologist, through the non-profit research organization Ocean Arks International of Falmouth, Massachusetts). Use the web address http://www.gaia.org/findhorn/ecovil/ecolm2.html for more information.

- *The GLOBE Program* is an international environmental science and education partnership, found in 4,000 schools (K–12) in 55 countries, for studying and understanding the global environment. Students make a core set of environmental observations at or near the school and report the data via the Internet to the *GLOBE* Student Archive. Student data sets posted on the World Wide Web are available for use by scientists and other students. Scientists are often available to provide feedback to students on their research. Some examples of categories for exploration include: atmosphere (clouds, temperature, precipitation), hydrology (surface water), soil (moisture, temperature, transect), and land cover/biology (qualitative and quantitative). The Internet address is http://www.globe.gov for the *GLOBE Program*.

TEACHING STRATEGY

If your school uses *ChemCom: Chemistry in the Community* (Kendall/Hunt Publishing Company), some students may be familiar with the chemistry of water and air. An environmental science teacher may also be a source of ideas, materials, and equipment.

- *Toyota TAPESTRY Grants for Teachers* are administered through the National Science Teachers Association (NSTA) and are offered to life science teachers (K–12) who propose innovative one-year programs to enhance science education in their schools. Many of the grants are for environmental education projects. Fifty grants of up to $10,000 each are awarded each year. The web address http://www.nsta.org provides more information and a description of recently awarded projects.

TEACHING STRATEGY

In order for students to write a field study report that meets your expectations, it will be important for you to explain to them about your requirements for the project. You will need to adjust the following scoring guide for a Level 4 report according to what project(s) your students have done and your own expectations.

EMBEDDED ASSESSMENT

The field study report for the long-term project may be used as an embedded assessment. The following rubric is provided to help you assess student ability to design a field study (if applicable), to demonstrate field skills, to write a report, and to work cooperatively with group members.

SCORING GUIDE FOR EVALUATION OF STUDENT RESPONSES

SCORE	DESCRIPTION
Level 4	Field study design is logical. Report is clear, well-organized, and contains all the components listed below. Student demonstrates technical field skills, the ability to relate concepts from the module to those in the field, and to work cooperatively in a group. Conclusions are logical and based on the data.
Level 3	Field study design is logical. Report is clear, well-organized, and contains most of the components listed below. Student demonstrates technical field skills, the ability to relate concepts from the module to those in the field, and to work cooperatively in a group. Conclusions are logical and are based on the data.
Level 2	Field study design has flaws. Report is less clear, not well-organized, relates some concepts to the field. Student may have difficulty working in a group. Conclusions do not follow logically from the data.
Level 1	Field study design has serious flaws. Report lacks organization and is missing many of the components listed below. Student does not act cooperatively in a group. Conclusions do not follow from the data.
Level 0	Field study and report are irrelevant or nonexistent.

A complete field study report includes clear and accurate presentation of the following:

- the topic being studied
- the question being asked
- the hypothesis (if applicable)
- the design of the field study (if applicable)
- the procedures (including materials)
- the data and observations
- the conclusions from the data
- concepts from the module related to the data and conclusions
- sources of possible error
- new questions as a result of the field study

Notes

See page iv in the Introduction section of this Teacher Guide for suggestions on how to use **scoring rubrics for assessment.**

Oh, What a Tangled Web

OVERVIEW In what ways do living things interact in a community? In the last learning experience, students discussed the roles organisms might play in their ecosystems. In this learning experience, they will look in greater detail into the role of a population of organisms in a community and focus on the relationships it has with other populations. In a biological community, organisms are dependent upon one another for survival. One of the most fundamental relationships among the biotic factors in an environment is based on nutrient and energy transfer through feeding. In this learning experience, students use a grocery store, its workers, and its products as an analogy for organisms linked by food relationships within a biological community. They then identify examples of food chains and examine a marsh ecosystem food web.

LEARNING OBJECTIVES

- **Students identify organisms and predict their ecological niches within their specific ecosystem.**

- **Students determine that organisms are dependent on one another and their environment for sources of food and energy.**

- **Students create food chains and a food web, then analyze how organisms in these pathways recycle nutrients and serve as a source of energy for populations within an ecosystem.**

▶ SUGGESTED TIME

- 3 class sessions (45–50 minute periods)

▶ MATERIALS NEEDED

For each group of four students:
- 1 large zippered plastic bag containing a sample of soil from one of three or four different locations (See Advance Preparation)
- 2 hand lenses or magnifying glasses
- 2 tweezers
- 2 dissecting needles (or coffee stirrers)
- 1 sheet of newspaper
- 1 petri dish
- dissecting microscope
- field guides (plants, insects, other invertebrates)
- 1 rotting log (if available)

For the class:
- 17 biota cards
- 1 cardboard "sun" (see Advance Preparation)

- 1 ball of yarn or string
- 1 scissors
- cellophane tape
- 1 meter stick or other pointer
- invertebrate field guides (optional)
- 5-cm diameter piece of pipe, small bulb planter, trowel, or spade

▶ ADVANCE PREPARATION

1. Prior to Session One:
   - collect core soil samples from several different undisturbed areas. Use a 5-cm diameter piece of pipe or a small bulb planter, or dig up one "scoop" of earth with a trowel or spade. Place in a large zippered plastic bag. Try to keep the layers of soil intact. Prepare one bag for each group of four students. Use soil samples from different locations, such as under a tree, in a sunny garden, or in a barren lot. Write the location from which each sample was taken on the bag.
   - assemble materials for the activity "Mystery Soil." Gather invertebrate field guides. If you want to take students to a pond, obtain permission for your class to go outside.
   - collect several small rotting logs and place them in a plastic bag (optional).

2. Prior to Session Three:
   - cut yarn or rope into 2-m lengths; cut six pieces for each student.
   - photocopy and cut the biota cards (found as Blackline Masters at the end of this learning experience) for use in the activity "What's for Lunch?"
   - create a cardboard "sun" for use in the food web activity "What's for Lunch?"

3. Order materials for Learning Experience 3—Round and Round They Go (see Learning Experience 3 Advance Preparation).

▶ TECHNOLOGY TOOLS

- *Return to Earth: Life Cycle of the Forest* by HRM Video focuses on various decomposers and how these organisms break down organic debris that accumulates in a forest. It is most useful at the end of Session Two, after the discussion of the organization of food chains, to emphasize the role or niche of decomposers in the recycling that occurs within environments. (See the Resource List in Appendix B at the end of this Teacher Guide.)

▶ ASSUMPTIONS OF PRIOR KNOWLEDGE AND SKILLS

- Students understand that plants make their own food, using the light energy from the sun and carbon dioxide from the air, and water from the soil in the process of photosynthesis.
- Students are aware that food provides organisms with energy as well as nutrients.
- Students are familiar with the characteristics of life.
- Students recognize that microorganisms are part of the environment.
- Students are familiar with concept-mapping.

▶ TEACHING SEQUENCE PREVIEW

SETTING THE CONTEXT

- In "Mystery Soil," students observe the community of organisms found in a soil sample and predict their ecological niches or roles in the community.

EXPERIMENTING AND INVESTIGATING

- Students analyze the relationship and dependencies of workers in a grocery store as an analogy for the complex interactions among species in an ecosystem.
- Students construct food chains and a food web and examine the effects of a "die-out" of a population.

PROCESSING FOR MEANING

- Students discuss the interrelationships of populations within an ecosystem.

Notes

Setting the Context
Session One

In this learning experience, students explore the ways in which organisms in an ecosystem interact with each other and with the environment, and examine how materials needed by these organisms are obtained and cycled through ecosystems.

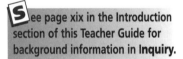

Prologue and "Mystery Soil": Student Manual pages 15–17.

Begin class by having students read the Prologue and the Introduction to the investigation "Mystery Soil." Divide the class into the same groups of four as in the previous learning experience. "Mystery Soil" is an exploratory exercise in which the students note the kinds and number of organisms found in the soil sample, and predict the possible role, or niche, of each type of organism.

See page xix in the Introduction section of this Teacher Guide for background information in **Inquiry**.

Tell students not to shake or mix the contents of their soil sample. Have groups carry out the Procedure. As students explore, circulate among the groups to observe student procedure and interactions. As students find different organisms, encourage them to use the field guides and to speculate on the roles these organisms play in their habitat.

TEACHING STRATEGY

Make available as many field guides as possible, so that students may identify their organisms. Students may need assistance in the use of field guides. It is not necessary for them to identify all the organisms they see. Most likely students will begin to connect the organism's niche to a supply of food and materials necessary for maintaining life.

TEACHING STRATEGY

To reinforce the concept of niche you may wish to have students analyze organisms found in pond water. If possible, take your students to a pond and collect samples from different depths. Have students analyze the organisms found at each depth and speculate as to their niches. One very effective way of analyzing this data is to have students create drawings of the organisms at each depth and then hang the drawings longitudinally in the same order. In this way, students can compare the types of organisms found at different depths and speculate as to how these organisms might interact and why they are found where they are.

Another alternative is to have students examine a rotting log. Pilot and field test teachers have reported that students were fascinated with the unfamiliar organisms found in that ecosystem.

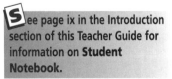

See page ix in the Introduction section of this Teacher Guide for information on **Student Notebook**.

▶ HOMEWORK

Have students write responses to the Analysis that follows "Mystery Soil" in their notebooks.

Session Two

Begin class by having students discuss the "Mystery Soil" investigation. This discussion should help students interpret what they have seen, speculate about soil as an indicator of and contributor to a rich diversity of life, and connect a population with its ecological niche or role in the community.

DISCUSSION QUESTIONS You may build on student observations by asking the following questions:

- Which soil location(s) had the most organisms? Why do you think that is so?
- Which had the least? Why do you think that might be?
- What other factors influence what lives in soil?
- What relationships exist between the plants and animals in this community?
- How might you contrast the organisms under the surface with those above the surface?
- What organisms might be present but unseen? What do they do?
- What effect might other living things in the area have on soil?
- What color were most of the animals you saw? Why might that be?

DISCUSSION QUESTIONS Continue exploring the concept of niche with student responses to the following questions (modified from the Student Manual):

- What are the relationships among the plants, the animals, and the soil in your soil community? *(Students might suggest that plants are a food resource for animals, some animals feed on other animals, and decomposers break down no-longer-living organisms. Animals that tunnel help to aerate the soil and also provide nutrients and minerals for a variety of organisms.)*
- What characteristics or factors did you use to determine the niche of each of the organisms? *(Students might respond that some characteristics would be what type of food-eater that organism is [producer or herbivore, etc.], its role within the ecosystem, and its interactions among other organisms within the ecosystem).*
- What do you think might happen if two populations of organisms with similar niches move into the same habitat? *(Students should speculate that one population will be successful while the other will not be.)*
- Describe the niches for the organism(s) in your bottle ecosystem.

Experimenting and Investigating

In order to introduce the complex network of interactions found in the food webs of a community of organisms, begin with the following challenge for the class. Have each group of four from the previous activity imagine the square meter of the soil ecosystem is analogous to a supermarket or grocery store and that each type of organism in the soil community was a different type of store worker. Write the following questions on the board or on chart paper:

- What types of workers are there in a grocery store?
- How do they interact with one another?
- Which can work independently?
- Which are dependent? In what ways?
- Are there other types of workers outside the store who are necessary to what goes on inside it?

Have students, in their groups, use these questions as they brainstorm the types of workers and their relationships and then create a chart, concept map, or diagram illustrating the interrelationships inside and outside the grocery store.

TEACHING STRATEGY

This grocery store analogy can be used to review the skill of concept-mapping to your students. After each group has listed the workers in a grocery store, have them draw lines to show any relationships. Have each group draw arrows and determine what action words or phrases they could write on the arrow line to connect the relationship. Once students have completed their map, you may want to describe specific concept map structures, which include a main concept, the ideas that come from the main concept, the ideas connected by arrows, and the short explanation of how they relate to one another on the connection line.

See page xii in the Introduction section of this Teacher Guide for information on **Concept-Mapping**.

Allow the groups 10–15 minutes to complete their maps or diagrams. When the groups have finished, have each group share its results with the class. If their maps or diagrams are limited to the visible store workers, ask them to think about the other people who supply the store with products. Students should include many of the following:

- farmers
- ranchers
- fishers
- shellfishers
- truckers
- buyers
- inspectors
- processors
- bakers
- harvesters

DISCUSSION QUESTIONS Discuss with the students the relationships and dependencies among the various workers. Ask students to imagine what would happen if any one or two of these types of workers disappeared. This question is intended to get students thinking about the consequences of an extinction event. The following questions might help frame this discussion:

- Which of the grocery store workers do you consider essential?
- What would happen if one group of those workers disappeared?
- Could other workers replace them (without changing their job description)?
- Which workers might not be essential to the store?
- What would happen if one of those groups of workers disappeared?
- Which of the groups would you consider essential for your survival? Explain your reasoning.

Students should recognize that the complete loss of any of these groups of people would have a direct impact on the grocery store. The loss of some of the groups of workers—stockers, truckers, cashiers, buyers—would destroy the ability of the store to function. Losing others, such as deli workers, fishers, ranchers, farmers, bakers, processors, and harvesters, would curtail choices.

DISCUSSION QUESTIONS Continue the discussion by having students compare their square-meter ecosystem or the ecocolumn ecosystem with the grocery store and think about the community of organisms they observed in their field sites. You may want to ask questions such as:

- How is the grocery store similar to an ecosystem?
- What do you think are the relationships among the organisms in your square meter? In your ecocolumn?
- Which of the organisms are essential to an ecosystem? What would happen if one of them disappeared?
- Would this niche be filled by another population? Explain your reasoning.

Allow 10 minutes at the end of the class session for students to think about the feeding relationships among organisms. Have each group construct a simple food chain having at least three links. Then have volunteers describe their food chains to the rest of the class. Write the food chains on the board or on chart paper. Have students specify which organisms are producers and which are consumers.

DISCUSSION QUESTIONS To help students analyze the organization of food chains, you may wish to ask questions such as the following:

- What do you notice about the sizes of the organisms in your food chain? What do you know about the number of each type of organism?
- What is the significance of the order of the organisms in a food chain?
- What pattern do you see among all the food chains?
- What unseen organisms are also part of the larger concept of food chains?

Decomposers may not have been mentioned, but students should be aware of their existence and of their role in the flow of matter through an ecosystem.

It is essential that students understand that the process of photosynthesis is the basis of the energy and nutrients in food chains and that all organisms obtain their food and energy from photosynthesizing organisms either directly or indirectly.

SCIENCE BACKGROUND

All energy comes from the sun; and only photosynthetic (or chemosynthetic) organisms, such as green plants, are able to use energy from the sun to convert inorganic compounds into sugar (with oxygen and water as byproducts). The light energy from the sun is changed to chemical energy, stored in such products as sugar. The process of photosynthesis is the basis for life on Earth. Plants, the first link in a food chain, serve as the nutrient and energy source for other organisms (herbivores and carnivores) in the chain.

▶ HOMEWORK

Inform students that they are to observe an organism for about 10 minutes (in a yard, a park, an aquarium, on a sidewalk, etc.) and describe, in an essay, its habitat, and its relationship to other organisms in the ecosystem.

Session Three

In this session, students explore how interconnecting food chains form a food web. Have students read the Introduction and follow the Procedure for Part A of "What's for Lunch?" In this activity, students connect organisms in an ecosystem based on their feeding relationship to create a food web. Distribute one biota card to each student until you run out. Clear a place on the floor or on a large table for students to create the food web. Place the cardboard sun in the middle of the area. Begin with

"What's for Lunch?":
Student Manual pages 17-19.

one of the plant species and have the student with that card tape a piece of yarn to connect it to the sun. Continue with only two or three students creating links at a time. Students without biota cards should help with the taping, the yarn, and finding the connections. Table 2.1 below is a table which shows all the feeding relationships in this simulation:

Table 2.1

BIOTA CONSUMERS	CONSUMER'S FOOD SOURCE
Little brown bat	Swamp milkweed leaf beetles Mosquitoes
Muskrat	Spring peepers Freshwater clams Great bulrushes Broad-leaved cattails
American black duck	Freshwater clams Broad-leaved cattails Spring peepers Fallfish
Marsh hawk	Spring peepers Ribbon snake Red-winged blackbird
Sora (wading bird)	Mosquitoes Freshwater clams Great bulrushes
Red-winged blackbird	Mosquitoes Broad-leaved cattails
Ribbon snake	Spring peepers Fish
Spring peeper (tree frog)	Swamp milkweed leaf beetles
Spotted turtle	Insects Freshwater clams Spring peepers
Fallfish	Mosquitoes
Freshwater clam	Algae Mosquitoes
Swamp milkweed leaf beetle	Swamp milkweed
Mosquito	Blood of mammals (female) Flower nectar (males)

Learning Experience 2   Oh, What a Tangled Web

Notes

Processing for Meaning

DISCUSSION QUESTIONS Have students step back and examine the completed food web, and have volunteers point to different food chains in it. You also may wish to ask questions such as the following:

- How many food chains do you think are needed for a food web? How few?
- How many interconnections do you think there are in this food web?
- In which direction is the flow of food or energy?
- How is this example similar to the grocery store analogy?
- What roles do bacteria and other decomposers play in this food web?

Before continuing to Part B of the activity, ask students to predict what would happen to the web if insects were to disappear, and to give reasons for their prediction. Then have students complete Part B. In "Local Extinction," students simulate the application of an insecticide to their food web. All mosquitoes and leaf beetles will be killed by such an application. Students observe which organisms are affected and are asked to describe the effect on the whole community.

DISCUSSION QUESTIONS The following questions may help guide this discussion:

- What was the first thing that happened after the insecticide was sprayed?
- What were the immediate consequences of the death of the mosquitoes and the beetles?
- Which other organisms were affected by the insecticide application? How and why?
- Which organisms were not affected? Why not?
- Is the disappearance of only one organism from an ecosystem possible? Why or why not?
- Why was this ecosystem so seriously affected by the insecticide?

▶ **HOMEWORK**

Have students write responses to the "What's for Lunch?" Analysis, then read "Managing Mosquitoes," an actual event in which the use of DDT for mosquito control had wide-ranging ramifications, and write responses to the Analysis.

ASSESSMENT

▶ *Do students understand that each population has an important role in a community and that removing organisms causes disruption in the ecosystem?*

"Managing Mosquitoes": Student Manual pages 19–20.

Learning Experience 2   Oh, What a Tangled Web

LITTLE BROWN BAT
Myotis lucifugus

DESCRIPTION

The little brown bat's wingspan is generally about 15–25 cm. It has a small, brown body and darker brown leathery wings. *Myotis,* like many other types of bats, feeds on insects which it scoops into its tail with wing movements while flying. Females give birth to a single offspring each year.

HABITAT

The little brown bat makes its home in caves and attics, hollow trees, quarries, and eaves of buildings.

MAJOR FOODS

mosquitoes, beetles

BIOTA CARD

MUSKRAT
Ondatra zibethica

DESCRIPTION

The muskrat is a medium-sized, dark brown rodent that spends much of its time swimming. It has webbed hind feet and a long, naked tail which it uses as a rudder when swimming. A muskrat's litter size can range from 1–8, and females have an average of three litters per year.

HABITAT

The muskrat makes its home in marshes, shallow parts of lakes and ponds, slow-moving streams, and drainage ditches, preferring areas with cattails. It may dig a den in the bank of a stream or ditch, or use weeds to make a dome-like nest chamber over shallow water to make a chamber.

MAJOR FOODS

cattails, bulrushes, spring peepers, freshwater clams

BIOTA CARD

AMERICAN BLACK DUCK
Anas rubripes

DESCRIPTION

The male black duck is blackish-brown overall with a paler head and face. The wings are white-lined, legs and bill are yellow, and feet are orange-red. The females are similar, but have duller colored legs and feet and a dark, mottled bill. They produce 6–11 eggs in one brood each year.

HABITAT

Black ducks are found and breed in the marshy borders of ponds, lakes, rivers and streams, wooded swamps, fresh, salt and brackish marshes and meadows. They winter in the open coastal and interior marshes.

MAJOR FOODS

freshwater clams, cattail seeds, spring peepers, fallfish

BIOTA CARD

MARSH HAWK
Circus cyaneus

DESCRIPTION

The marsh hawk is common in open country areas. Typically about 55 cm long, the male has gray upper parts with black-tipped wings, a white underbelly, and a white rump. The females have brown upper parts, brown-streaked buff under parts and a white rump. Immature birds are similar to females, except for rich cinnamon-colored under parts. Females lay their eggs in clutches of 3–9 eggs in a single brood per year. Hawks are easy to recognize by their in-flight behavior; these hawks spend most of their time gliding with their long, narrow wings held in a shallow V-shape, only occasionally flapping their wings to maintain flight.

HABITAT

The marsh hawk is found in open country, fresh and salt water marshes, swamps and bogs, and wet meadows.

MAJOR FOODS

spring peepers, red-winged blackbirds, snakes, small rodents

BIOTA CARD

Sora (Wading Bird)
Porzana carolina

DESCRIPTION

The sora is a medium-sized rail (wading bird), about 20–25 cm long, with a short bill that resembles that of a chicken. It is a shy creature and spends much of its time hiding in marsh vegetation. Soras are weak fliers, but strong swimmers. Adults have olive-brown upper parts streaked with broad black bands and fine white lines. Their under parts are striped with black, white and gray. Their faces and throats are black, their bills yellow. The sora lays between 6–13 eggs in one brood each year.

HABITAT

The breeding sora lives in freshwater marshes, ponds, bogs, and wet grassy meadows. It prefers areas with dense stands of cattails in deep water and mud.

MAJOR FOODS

mosquitoes, bulrush seeds, freshwater clams

BIOTA CARD

Red-Winged Blackbird
Agelaius phoeniceus

DESCRIPTION

The red-winged blackbird is robin-sized (about 23 cm long). The males are black with a bright red and yellow shoulder patch, while females and immature birds are a light brown that is heavily streaked with dark brown. The females usually lay between 3–5 eggs in one brood and have two or three broods of chicks each year.

HABITAT

Red-winged blackbirds breed in swamps, marshy areas, wet meadows, and ponds, usually near dry fields. They prefer wetlands with abundant growth of cattails, bulrushes, sedges, and reeds.

MAJOR FOODS

mosquitoes, cattail seeds

BIOTA CARD

Ribbon Snake
Thamnophis sauritus

DESCRIPTION

Ribbon snakes are generally 45–65 cm long. They are sleek garter snakes and are darkly colored with three stripes that are generally bright yellow, but also can be orange or greenish. Their bellies are generally a dull yellow color. Ribbon snakes seldom stray far from swamps, pools, marshes or bogs. They are viviparous, meaning that they give birth to live young, as opposed to eggs. Ribbon snakes give birth to 3–20 young in each yearly laying.

HABITAT

The ribbon snake is semi-aquatic. It lives on the edges of streams, in swampy areas, wet meadows, ponds, bogs, and ditches. It prefers areas with brushy vegetation at the water's edge, where it can hide itself from predators.

MAJOR FOODS

frogs, including spring peepers; worms, fish

BIOTA CARD

Spring Peeper (Tree Frog)
Hyla crucifer

DESCRIPTION

Spring peepers are small (1.5–3 cm) tree frogs. They are olive green, brown or gray in color and have a darker X shape on their backs. Peepers vocalize in the spring, making a high, piping note that they repeat at intervals of about one second. They lay eggs singly in batches of 800–1,000 eggs near the bottom of shallow, weedy ponds and attach their eggs to submerged plant stems.

HABITAT

The spring peeper is found in woodlands where there is much brushy undergrowth, or in woodlots near ponds, swamps, and marshes. It prefers to climb into trees and shrubs that are in or near standing water.

MAJOR FOODS

beetles

BIOTA CARD

SPOTTED TURTLE
Clemmys guttata

DESCRIPTION

The spotted turtle is usually 8–12 cm long. It has a dark brown or black smooth carapace (shell) and head, both with small yellow spots. The number of spots increases with age. Females lay one to four eggs in the soil of marshy pastures.

HABITAT

The spotted turtle lives in small, shallow bodies of water, like woodland streams, wet meadows, bogs, small ponds, marshes, swamps, roadside ditches and brackish tidal creeks. It hibernates in the muddy bottoms of bogs and marshes.

MAJOR FOODS

insects, freshwater clams, spring peepers

BIOTA CARD

FALLFISH
Semotilus corporalis

DESCRIPTION

The fallfish is one of a group of fish known as minnows. It has large scales, large eyes, and a compressed body. Its snout is long and rounded, slightly overhanging its large mouth. The scales on the back and upper sides of the adult fish are outlined in black. The young have a black stripe along their sides. During spring, fallfish dig pits in stream beds and bring in pebbles and small rocks where they lay their eggs.

HABITAT

Fallfish are found in gravel and rubble-bottomed pools, in runs of small to medium rivers, and at lake margins.

MAJOR FOODS

mosquito larvae

BIOTA CARD

Freshwater Clam
Sphaerium sp.

DESCRIPTION

The freshwater clam is a tiny clam with a translucent shell (up to 1-cm diameter) that makes the clam glitter below the water's surface. The shells have prominent ridges or growth rings. The fertilized eggs are incubated in a specialized chamber inside the adult female. Before the end of summer, the female releases the young clams, which drop to the water's bottom and continue to develop.

HABITAT

The freshwater clam lives in swamps and marshlands.

MAJOR FOODS

algae, mosquito larvae

BIOTA CARD

Swamp Milkweed Leaf Beetle
Labioderma clivicollis

DESCRIPTION

Swamp milkweed leaf beetles are 0.5–1.0 cm long. The insects are oval and bluish or greenish black. They have orange/yellow elytra (forewings) with two black marks forming an X across the midline of the beetle. The beetle lays long, yellow eggs on milkweed leaves. The eggs soon hatch into larvae, which feed on the leaves and eventually drop to the ground and pupate there. The adults emerge in late summer.

HABITAT

The swamp milkweed leaf beetle lives in marshes and at the edges of streams.

MAJOR FOODS

swamp milkweed leaves

BIOTA CARD

MOSQUITO
Culex sp.

DESCRIPTION

The mosquito is a small fly in the order Diptera. Males do not bite; females do. Some types of mosquitoes transmit infection (including malaria, yellow fever, dengue, etc.) by their bite. Females lay several hundred eggs in masses on the surface of standing water. The eggs hatch into larvae in a few days, then pupate in about a week to emerge soon after as adults.

HABITAT

Most moist areas support mosquito life, and the insects are especially abundant around marshes, ponds, pools, cavities in trees, and other places where water collects.

MAJOR FOODS

Females: blood of mammals (muskrat, little brown bat)
Males: nectar of swamp milkweed flowers

BIOTA CARD

ALGAE

DESCRIPTION

Algae are widely ranging photosynthetic organisms which can be unicellular or multicellular, from the tiny (about 0.01-cm) diatoms to the giant kelp which can grow to 35 meters or longer. All algae contain chlorophyll and are adapted to aquatic life. They are generally considered plants, but have none of the stems, roots, or leaves normally associated with plants, although they may have leaflike structures. Algae may have periods of "blooming" during which they multiply greatly. These blooms occur during the warmer seasons or when there is excess nitrogen phosphores and other nutrients in the water.

HABITAT

Depending on the species of algae, they can be found in fresh water or sea water.

BIOTA CARD

Broad-leaved Cattail
Typha latifolia

DESCRIPTION

The broad-leaved cattail is a medium to tall perennial (plant that has a life span of more than two years), that can grow to up to three meters tall. It has simple, sheath-like leaves that ascend along the stem of the plant, alternating from side to side. Its dark brown terminal spike (the "cat tail") looks as if it is covered with velvet. In addition, cattails have small terminal flowers. They also have tuberous roots that send up new shoots each year, allowing the root base to spread, and mud to build up around the base of the plant.

HABITAT

The cattail is found in freshwater marshes, ponds, and roadside drainage ditches.

BIOTA CARD

Great Bulrush
Scirpus validus

DESCRIPTION

The great bulrush is a tall, marshland reed. It grows up to 2.5 meters in height. It is characterized by a round, green stem, a single leaf, and loose clusters of green flower spikes. The bulrush grows in thick stands in one or more meters of water.

HABITAT

The great bulrush is found in the mud of shallow, fresh or brackish marshes.

BIOTA CARD

Swamp Milkweed
Asclepias incarnata

DESCRIPTION

The swamp milkweed is a one-meter tall plant with long, narrow, smooth, lance-shaped leaves. Its dull pink flowers are visited for the nectar by many kinds of insects. The milkweed has long, pointed seedpods filled with white tufted seeds that are scattered by the wind after the seedpod opens. When broken, the flesh of a milkweed oozes a thick, milky white juice.

HABITAT

The swamp milkweed is found in swamps and other wet areas.

BIOTA CARD

Round and Round They Go

OVERVIEW Sustaining life in an ecosystem requires a steady supply of the resources that organisms need to carry out fundamental life processes such as growth, movement, reproduction, photosynthesis, and respiration. In the last learning experience, students explored nutrients moving through food chains and food webs. In this learning experience, students explore several essential resources, including nutrients, energy, carbon, oxygen, nitrogen, and water.

Students are introduced to the concept of trophic levels and calculate the diminishing amount of energy available in each succeeding trophic level. They then set up an investigation to trace the cycling of carbon and, by inference, oxygen between *Elodea* and a snail in a closed system. Finally, they examine biogeochemical cycles and determine the importance of the carbon–oxygen, nitrogen, and water cycles to the maintenance of life on Earth.

▶ SUGGESTED TIME

- 4 class sessions (45–50 minute periods)

▶ MATERIALS NEEDED

For each group of three students:
- 3 sheets of unlined white paper (11 x 14 inches, legal size)
- felt-tip markers or colored pencils

For each group of four students:
- 4 pairs of safety goggles
- 4 culture tubes or test tubes with tight-fitting covers or stoppers
- 1 test tube rack
- 1 wax marking pencil
- 100–200 mL distilled water (or "aged" tap water)
- 1 eyedropper

LEARNING OBJECTIVES

- Students calculate the flow of energy through a biotic community.

- Students investigate the carbon–oxygen pathway in a closed living system.

- Students trace the pathways of carbon, oxygen, nitrogen, and water as they cycle among living organisms and the environment.

- Students examine the interconnections and interdependencies of the biogeochemical cycles and determine that the continued functioning of each cycle depends on the other cycles.

- 1–2 mL 0.1% bromothymol blue solution (aqueous)
- 4 small water (pond) snails
- 2 sprigs of *Elodea*
- access to a fluorescent lamp

For the class:
- 4 transparency sheets
- water cycle demonstration set-up (optional)
  - salt
  - hot water
  - 1 plastic bag
  - 1 spoon
  - ice cubes
  - 1 2-L plastic bottle
  - 1 1-L plastic bottle
  - 2 laboratory stands
  - 2 bowls

▶ ASSUMPTIONS OF PRIOR KNOWLEDGE AND SKILLS

- Students are able to use percentages in calculations.
- Students are aware that calories are a measure of the amount of energy released when food is oxidized.
- Students understand the concept of a closed system.
- Students understand the use of a control in an experiment.
- Students are familiar with chemical indicators.
- Students are aware that inorganic elements in the environment provide organisms with the materials they need to maintain life.

▶ ADVANCE PREPARATION

1. Prior to Session One, order materials. Water (pond) snails and *Elodea* may be ordered from a biological supplier or they may be purchased from an aquarium supply store.

2. Prior to Session Two, prepare a 0.1% bromothymol blue aqueous solution by mixing the powder form with water. (***NOTE:*** Alcohol, found in many commercial preparations, may damage or even kill the organisms.)
   - Prepare a 0.1% aqueous solution by dissolving 0.5 g bromothymol blue powder in 500 mL distilled water. If solution is not blue, the water used must be somewhat acidic (below pH 6). In this event, a 0.4% solution of sodium hydroxide may be added, one drop at a time, until the water turns blue.

SAFETY NOTE: Bromothymol blue and sodium hydroxide are both skin and eye irritants. Wear safety goggles, gloves, and a protective apron during the preparation. If contact occurs, flush the affected area with water immediately.

Learning Experience 3 Round and Round They Go

3. Prior to Session Three, set up four tubes as instructed in "It's Elemental" (found on page 26 in the Student Manual) and place them in the dark. These tubes will serve as a comparison with student tubes kept in the light.

4. Prior to Session Four, prepare four transparencies: one each of the "Carbon-Oxygen Cycle," the "Water Cycle," and the "Nitrogen Cycle" found as Blackline Masters at the end of this learning experience; and one of "Overlaps among the Cycles" found on page 32 in the Student Manual.

5. Make copies of the Blackline Masters "Carbon-Oxygen Cycle," "Water Cycle," and "Nitrogen Cycle" found at the end of this learning experience. Copy one set for each student.

▶ TECHNOLOGY TOOLS

- *Ecosystems* is a video produced by Hawkhill Science. Part 2 outlines several major concepts used by ecologists to study ecosystems. Topics such as thermodynamics, energy flow, the cycling of matter, food chains, and the producer-consumer-decomposer cycle are covered. This video complements the readings in this learning experience and the observations that students may be making in their model ecosystem. (See the Resource List in Appendix B at the end of this Teacher Guide.)

▶ TEACHING SEQUENCE PREVIEW

SETTING THE CONTEXT
- Students use data to calculate the amount of energy loss in each succeeding trophic level and discuss the implications.

EXPERIMENTING AND INVESTIGATING
- Students set up a closed system and trace the carbon–oxygen pathway between Elodea and a snail.
- Students explore and draw one of the biogeochemical cycles, carbon–oxygen, nitrogen, or water, and explain it to the group.

PROCESSING FOR MEANING
- Students examine the connections among biotic and abiotic factors in the environment and the ramifications of having finite amounts of materials on Earth.

APPLYING
- Students analyze how each biogeochemical cycle is dependent on the functioning of the other cycles.

Notes

TEACHING SEQUENCE

Setting the Context

Session One

Begin class by having students discuss their responses to the Analysis in "What's for Lunch." Students should recognize that the connections among the organisms follow general patterns: producer to consumer, smaller size to larger size, etc. Continue the discussion by having several volunteers explain the flow charts that they did for homework at the end of Learning Experience 2. You may wish to use their explanations to draw on the board a flow chart that shows the result of the spraying of DDT in Borneo. Continue by discussing students' responses to the Analysis. During this discussion, it is important that students mention specific organisms and specific effects, and not speak of ecosystem imbalance in general terms only.

Divide the class in pairs and have them read the Prologue and the Introduction to "The Flow of Energy." In this Setting the Context activity, students calculate the energy flow through a four-link food chain and analyze the implications of energy loss on the number and size of organisms found in each trophic level. Ask students to examine Table 3.1 and determine how they would calculate the amount of energy available to a herbivore which consumes the primary producer.

TEACHING STRATEGY

Students should suggest combining the 60% loss from metabolism and the 20% loss from waste and calculate that 80% (or 0.8) of 20,000 kilocalories (kcal) is lost. If students are having difficulty with the calculation, allow a few moments for students to pair up and work it out.

Have students carry out the Task. In step 2 of the Task, students are asked to present their calculations of the energy loss in graph form. This representation of numbers in a different form is intended to see if students understand the meaning of the numbers they have reached. It is not necessary that the graph show interim numbers, but rather that the trend of energy loss through trophic levels is shown.

TEACHING STRATEGY

Circulate among the pairs to see if they are using the right numbers in their calculations. The mathematics involved is straightforward. Eighty percent (or 0.8) of 20,000 kcal is 16,000 kcal used by the producers which leaves 4,000 kcal available for the herbivore. The herbivore uses 85% of the 4,000 available kcal in its metabolism and waste, and thus only 600 kcal are available for the primary carnivore. Using Table 3.1 in the Student Manual, have students continue the calculations for each trophic level using the appropri-

---

ASSESSMENT

▶ *Do students understand the results and the implications of destroying one or more organisms in a food web?*

Prologue and "The Flow of Energy": Student Manual pages 23–25.

See page xx in the Introduction section of this Teacher Guide for background information on **Critical Thinking.**

Learning Experience 3   Round and Round They Go

ate energy loss percentages. If pairs are having difficulty visualizing how to set up the graph, you may wish to suggest that they place the trophic levels on the x axis and the kilocalories on the y-axis.

SCIENCE BACKGROUND

Estimates of energy loss in living systems vary widely. The most widely quoted estimate by ecologists is called the "Rule of 10." It states that, on average, only about 10 percent of the energy fixed by plants is ultimately passed on to herbivores. Only 10 percent of the energy herbivores accumulate is passed on to the carnivores that eat them. And only 10 percent of that energy is transferred to carnivores on the third trophic level. This paints an extremely inefficient picture of energy transactions in nature.

DISCUSSION QUESTIONS

Leave time at the end of the session to discuss the activity. When students have finished, initiate a discussion with the following (modified from the Student Manual):

- How much of the original 20,000 kcal of energy is available for the secondary carnivore? For a tertiary carnivore? *(Students should respond that 30 kcal are available for the secondary carnivore, but only 1.5 kcal for a tertiary carnivore.)*

- Explain the flow of energy through trophic levels. *(Students should respond that energy is transferred and lost through each trophic level and that the flow is unidirectional from producer through consumers.)*

- Is there more energy transferred to the consumer after eating a pound of rice or a pound of meat? Explain your response. *(Students should explain that a pound of rice transfers more energy since rice is a producer and is the lowest trophic level.)*

- Explain the ecological principles illustrated by Figure 3.2. *(Students should state that each of the blocks represents a trophic level; that there are more producers than herbivores, more herbivores than primary carnivores, etc.)*

- How many trophic levels did you see in the Antarctic food web? Give an example of each level.

ASSESSMENT

▶ Do students understand that energy has been used by organisms on each trophic level and is not "lost" but is no longer available to organisms in the next trophic level?

SCIENCE BACKGROUND

The laws of thermodynamics can be applied to ecosystems. The first law states that the total amount of energy in a system remains constant, even though its form may change. The second law asserts that, while energy is not lost, its capacity to do work is decreased. Energy in a feeding relationship is passed from organism to organism and utilized at each level. Each organism in a food chain changes the energy it takes in as it undergoes the processes of digestion, respiration, and excretion. Energy is changed into a different form and is unusable to the organism at the next trophic level. Thus, the flow of energy is unidirectional through the trophic levels. Energy is constantly being replaced in an ecosystem by sunlight and plant photosynthetic activity.

▶ HOMEWORK

Have students create an "ecosystem" concept map with the following terms: biotic factors, environment, habitat, community, abiotic factors, niche, food chain, population, producer, consumer, decomposer, food web, photosynthesis, trophic level, energy flow.

Remind students to connect the terms with arrows and to use linking words that explain the connection on the arrow.

▶ EMBEDDED ASSESSMENT

Student concept maps may be used as an embedded assessment. The following rubric is provided to help you assess student understanding of the concepts in this module and the complex interactions within an ecosystem.

▶ SCORING GUIDE FOR EVALUATION OF STUDENT RESPONSES

SCORE	DESCRIPTION
Level 4	Student concept map is clear and contains all the terms listed. Map is organized from general to specific, has branches and is not linear; shows logical appropriate relationships among concepts, and uses appropriate linking words.
Level 3	Student concept map is clear and contains all the terms listed. Map shows logical and appropriate relations with a few minor errors and uses logical linking words.
Level 2	Student concept map is less clear and contains most of the terms listed. Map is fairly linear, not clearly developed, and shows flaws.
Level 1	Student concept map contains some of the terms listed, demonstrates little understanding of the relationships among terms, and contains major flaws.
Level 0	Map is totally incorrect or nonexistent.

An example of a Level 4 concept map is provided on page 43.

See page iv in the Introduction section of this Teacher Guide for suggestions on how to use **scoring rubrics for assessment.**

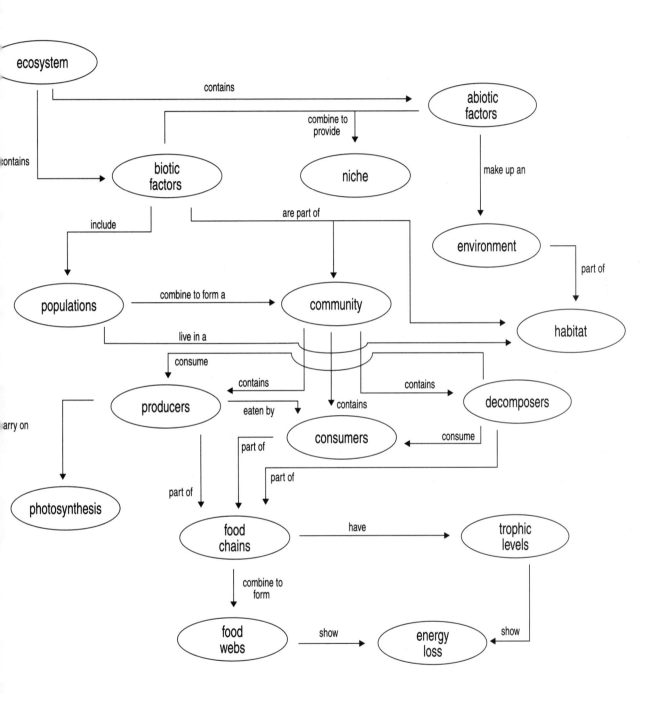

Learning Experience 3 Round and Round They Go

Experimenting and Investigating

Session Two

Begin class by having one or two volunteers present their concept maps. Encourage other students to comment on the maps and suggest other connections.

> "It's Elemental":
> Student Manual pages 26–27.

Have students read the Introduction to "It's Elemental" and carry out the Procedure. This investigation provides students the opportunity to investigate the carbon–oxygen pathway in a closed system. Inform students that they will be observing the tubes from the experimental setup every session for at least one week and recording their observations in their notebooks. At the end of one to two weeks, or when you think no new information can be obtained, have students complete the Analysis. When students are recording their observations, you should observe and record your own observations of the test tubes you have set up and placed in the dark (see Advance Preparation).

SCIENCE BACKGROUND

The reversible chemical equation for the formation and dissociation of carbonic acid is

$$CO_2 + H_2O \longleftrightarrow H_2CO_3.$$

MODULE CONNECTION

If students have completed the *Insights in Biology* module *The Matter of Life*, remind them that they have seen part of the carbon–oxygen cycle before. In *The Matter of Life*, students carry out an experiment showing changes in the level of carbon dioxide in the air in the presence of living organisms. An indicator is used to detect the presence of carbonic acid (carbon dioxide dissolved in water).

A question in the Analysis asks students to predict what would happen if the tubes were placed in the dark. After they discuss their responses, show your set of tubes to the class and have them comment on the difference.

TEACHING STRATEGY

As an alternative to your setting up a set of tubes in the dark, you may wish to have half of the student groups place their tubes in the dark. If you choose to do this, have students compare the results, including the time factor, in each location.

The results of the investigation "It's Elemental" will be processed in Learning Experience 4.

Processing for Meaning
Session Three

DISCUSSION QUESTIONS In this session, students begin by discussing non-food resources that are necessary in order for organisms to maintain the characteristics of life. Begin the session by asking questions such as the following:

- What resources other than food are important for life?
- How long might a mammal such as a mouse survive without food? water? air?
- What makes Earth unique in its ability to sustain life?
- Is there an infinite supply of resources on Earth? If not, what conditions are necessary for a continuous supply of resources?

MODULE CONNECTION

If students have completed the *Insights in Biology* module *The Matter of Life*, they should be familiar with the ways in which life processes are dependent upon the availability of resources. You may wish to discuss how studying the survival of an individual organism is similar to and different from studying the survival of organisms in an ecosystem.

Divide the class into groups of three and have them read the Introduction to "An Infinite Loop." This activity is planned as a jigsaw learning in which one student from each group, with one student from the other groups, will analyze the description of one of the biogeochemical cycles. Assign students to a cycle and have them carry out the Task. This should result in 3 large groups. (See the Teaching Strategy that follows). Each student will draw a representation of the cycle. They should complete this part of the Task in 20 minutes. When they have finished, students should return to their original groups of three and explain their cycle to the other two members. Collect the student cycle drawings. You may wish to post a representative sample in the classroom.

"An Infinite Loop":
Student Manual pages 28–32.

ASSESSMENT

- Has each student shown the cycles as cycles?
- Are the relationships between the biotic and abiotic factors clear and accurate?

TEACHING STRATEGY

You may set up the jigsaw in the following way: Assign student 1 in each group to the carbon–oxygen cycle; student 2 to the water cycle; student 3 to the nitrogen cycle. The latter cycle is the most difficult and you may wish to adjust the assignment accordingly. Have all the students studying the carbon–oxygen cycle assemble in one area of the room where, in pairs, they will assist each other in interpreting the descriptions. Do the same with the other two large groups. When students indicate they understand the cycle, hand out the 11 x14 inch paper and markers or colored pencils so that they each may draw the cycle.

TEACHING STRATEGY

You may wish to set up a demonstration of the water cycle using the following: table salt, hot water, plastic bag to form a drape, spoon, ice cubes, 2-L plastic bottle cut in half lengthwise, 1-L plastic bottle with hole on its long side, laboratory stands, and two bowls (one small and one larger). Place 2–3 tbsp of salt in the larger bowl, add a small amount of water and stir. Tilt up the 2-L cut bottle on its side on the stands and place the 1-L bottle inside it. Add ice to the 1-L bottle. Place the small bowl under the 2-L bottle top. Arrange the plastic so that it hangs over the small bowl. Add hot water to the dissolved salt in the larger bowl and place it under the 2-L bottle. Observe over the course of the class session. Taste the collected water in the small bowl.

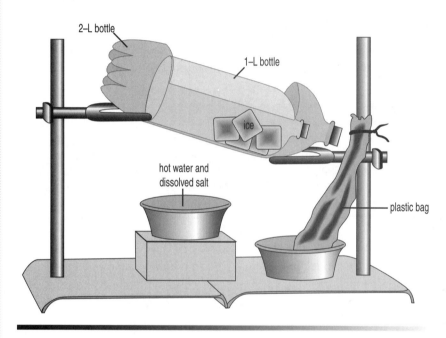

Figure 3.1
Setup for demonstrating the water cycle.

▶ **HOMEWORK**

Have students write and/or draw responses to the Analysis that follows "An Infinite Loop."

Applying

Session Four

DISCUSSION QUESTIONS Begin this session by having volunteers demonstrate and explain to the class the cycles they depicted for homework. Begin a discussion by asking the following (modified from the Student Manual):

♦ When you drew separate cycles for carbon and oxygen, at what points were the cycles unlinked? *(Students should realize that when* *either element is part of a compound that doesn't include the other, such as H_2O or CH_4 [methane], the cycles are separate.)*

- What biotic or abiotic factor did you remove from the carbon–oxygen cycle? What was the result? From the nitrogen cycle? From the water cycle? *(Student responses should cover a variety of factors, with the corresponding results mentioned.)*
- How did you interpret the quotation by Kamo no Chomei? *(Students should state that ocean water is constantly being replaced and renewed by precipitation, runoff, and the streaming from ground water and the water table).*
- All of the elements in these cycles are finite or limited. How is this fact important in your thinking about cycles? *(Student responses should indicate that since there are finite amounts of these elements, recycling is essential without which cycles would stop and life could not continue).*

Distribute the copies of the cycles to the students to be used as a resource (see Advance Preparation). If you decide to use the transparencies of the cycles as a review, have a volunteer "walk through" each cycle.

Allow 10 minutes for students to examine Figure 3.4 in the Student Manual. Ask them to determine how the three cycles overlap and interconnect with each other. Place the transparency of this diagram on an overhead as students work.

DISCUSSION QUESTIONS

Gather the class together to discuss the interconnections and interdependencies of each cycle. In this discussion, students will use their knowledge of the cycles to build an understanding of the ways in which each cycle depends, in part, on the functioning of the other cycles. The overlap is found within the geological components as one cycle draws on materials from another cycle. Students should, then, be able to extend their understanding to think of the planet as an ecosystem, one that is in a state of dynamic equilibrium. Questions such as the following may help you to facilitate the discussion:

- Where do these three cycles overlap on Earth? *(Students should respond that they overlap in the atmosphere, the soil, the oceans, and in organisms.)*
- What are some examples of where one cycle is dependent on the byproduct of another cycle? *(Students should respond, for example, that the nitrogen cycle uses oxygen when nitrites are converted to nitrates; the carbon–oxygen cycle depends on water and nitrates; and so on.)*

ASSESSMENT

▶ **Do students understand the significance of the cycles to life on Earth?**

▶ **Do students understand that the cycles depend on natural biological, chemical, and geological processes?**

- How is ecosystem Earth dependent on the flow of these elements? *(All plants and animals require these elements to sustain life.)*
- If one cycle ceased to function, would the other two cycles be affected? If so, in what ways?
- How would you defend the statement that the "balance" on Earth is actually one of dynamic equilibrium?

▶ **HOMEWORK**

If you feel that no further information can be obtained from observing the ecocolumns, have students write a laboratory report for "Life in a Jar" and be prepared to discuss their conclusions in class. (*NOTE:* Return to the Applying section of Learning Experience 1 to conclude the investigation.)

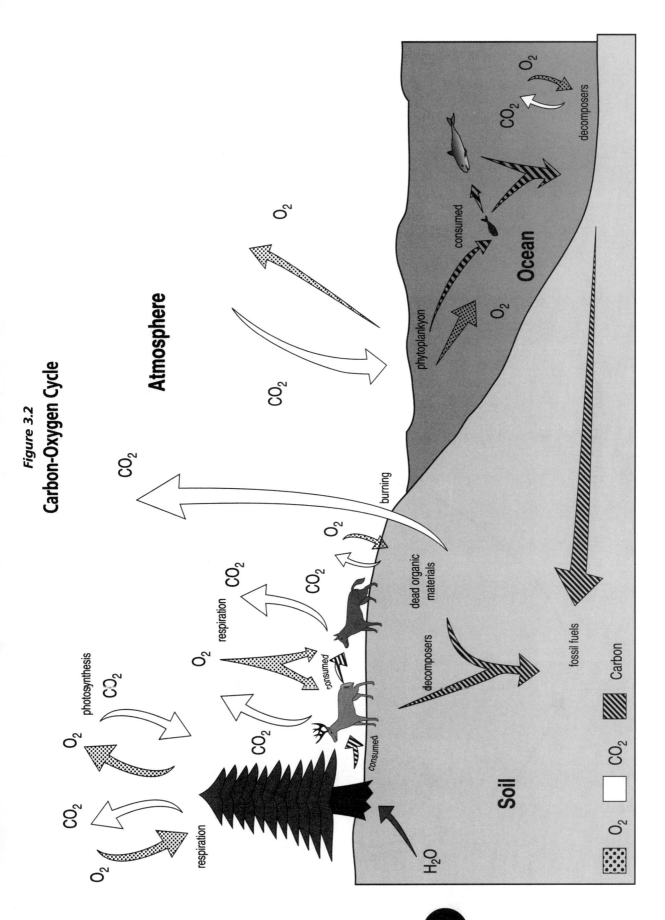

Figure 3.2
Carbon-Oxygen Cycle

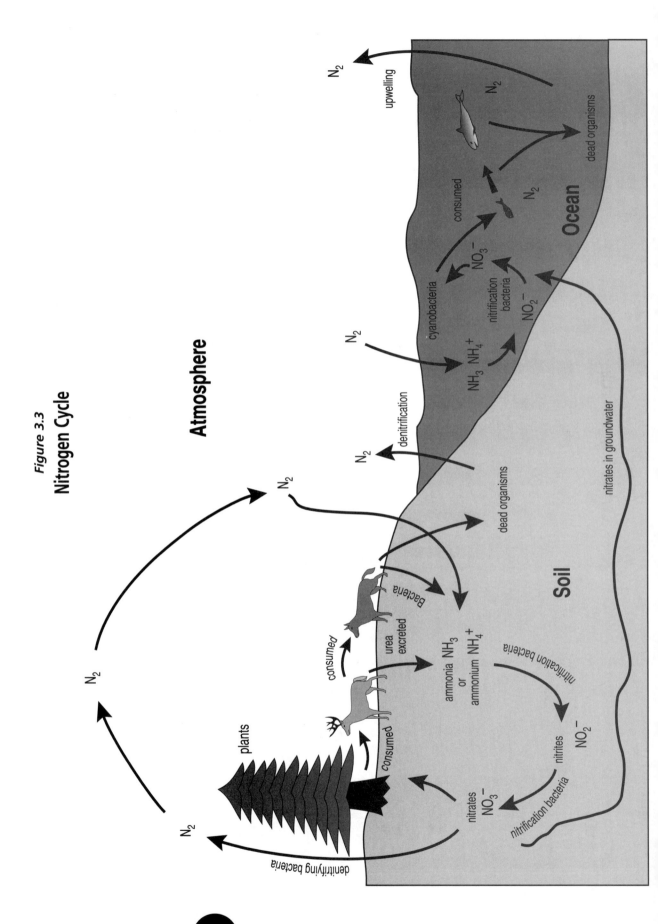

Figure 3.3
Nitrogen Cycle

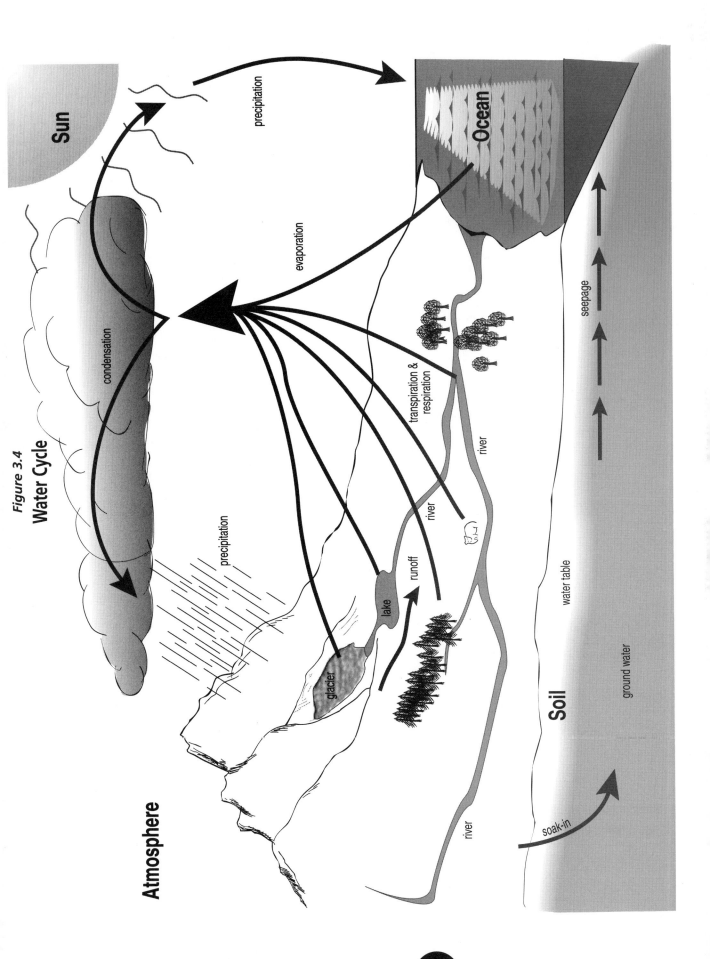

Figure 3.4
Water Cycle

POPULATION PRESSURES

OVERVIEW What underlying factors cause natural populations of organisms in an ecosystem to increase or decrease? What patterns can be discerned and what is the significance of these patterns? In this learning experience, students examine the factors that influence the growth or reduction of populations in an ecosystem over time. These interplaying factors help populations maintain relatively consistent numbers.

Students begin the study of population dynamics by exploring several different growth patterns: the exponential or J-shaped curve, in which populations with ample resources increase through rapid reproduction; the "boom and crash" curve, in which an isolated population increases rapidly and then decreases quickly; and the sustaining or S-curve in which the population grows rapidly for a time and then levels off. Students then investigate the interrelationships among populations within an environment by carrying out a simulation of predator-prey relationships and reading a case study about the interactions among several populations—mice, deer ticks, gypsy moths, and acorns—in a northern forest.

▶ SUGGESTED TIME

- 4 class sessions (45–50 minute periods)
  - Session Four should occur approximately one week after the completion of Session Three.

▶ MATERIALS NEEDED

For each pair of students:
- 1 sheet of graph paper

For each group of four students:
- 100 hare cards (5 cm x 5 cm) (See Advance Preparation)
- 25 lynx cards (10 cm x 10 cm) (See Advance Preparation)
- 1 ruler or tape measure
- masking tape
- 1 sheet of graph paper

LEARNING OBJECTIVES

▶ Students explore several growth curves and identify environmental factors that affect population growth.

▶ Students simulate the effects of predation on population size and determine the effects of changes in each population.

▶ Students analyze information on the populations of organisms that are interconnected in a northeastern forest ecosystem, and assess the factors that influence population size and, in turn, cause both Lyme disease and outbreaks of gypsy moths.

▶ ADVANCE PREPARATION

1. Prior to Session Two:
   - cut 100 hare cards (5 cm x 5 cm squares) from card stock or heavy paper for each group of four students.
   - cut 25 lynx cards (10 cm x 10 cm squares) from card stock or heavy paper for each group of four students.
2. Prior to Session Three, make one copy of the non-text illustration "Acorn Rollercoaster" for each student, found on page 62 of this learning experience.
3. Read the *For Further Study* "Be Fruitful and Multiply?" found at the end of this Learning Experience. If you choose to have students conduct the *For Further Study* read through the Advance Preparation and order materials if necessary.

▶ TECHNOLOGY TOOLS

- *Biology Explorer: Population Ecology* simulates the growth and interactions of populations. It complements this module by extending student experiences and allowing them to manipulate variables.
   The following concepts are explored in *Biology Explorer: Population Ecology:*
  - food relationships including food chains, food webs, trophic levels, and biomass pyramid
  - population growth including exponential growth, logistical growth, carrying capacity, and biotic relationships
  - limiting factors to population growth
  - niche and habitat
  - adaptations including food utilization and abiotic factors
  - management including ecosystem stabilization, biological control, and ecological decision-making

   You may wish to have students use *Biology Explorer: Population Ecology* several times during the module. If used following the "Lynx and Hare" activity, it will help students develop a deeper understanding of the variables that influence the population fluctuations that they have observed.
   The complexity and flexibility of this simulation may present some difficulties for your students. Students will need to pay attention to various tools and graphs and be prepared to spend time thinking and talking about what they are seeing and learning. See the Resource List in Appendix B of this Teacher Guide for more information about the software program.

▶ ASSUMPTION OF PRIOR KNOWLEDGE AND SKILLS

- Students are able to create a data chart and construct a graph.

> **FOR FURTHER STUDY**
> In the *For Further Study* "Be Fruitful and Multiply?" students trace and describe human population growth and begin to explore what that data means with respect to Earth's carrying capacity. Students interpret a graph of human population change over time, discuss factors that contribute to the continuing growth curve and think about the environmental impact of such growth.

See page viii in the Introduction section of this Teacher Guide for information on and rationale for **For Further Study.**

▶ TEACHING SEQUENCE PREVIEW

SETTING THE CONTEXT
- Students predict the factors that influence population growth, and graph the change in numbers over time of a yeast colony.

EXPERIMENTING AND INVESTIGATING
- In the activity "The Lynx and the Hare" students carry out a predator-prey simulation, record data, and analyze and graph the results.

PROCESSING FOR MEANING
- Students display their graphs and discuss the causes and effects of changes in the hare/lynx population.
- Students discuss the results of the "It's Elemental" investigation found in Learning Experience 3.

APPLYING
- Students create a poster showing the cyclic interrelations of five populations in a forest ecosystem.

Setting the Context
Session One

DISCUSSION QUESTIONS The purpose of this Setting the Context is for students to explore the growth of populations under both ideal and limiting conditions, to plot their data, and to think about the factors that determine each resulting population curve. Begin this session by having the class read the Prologue; then ask:

Prologue: Student Manual page 35.

- What resources would a male and female mouse in a cage need to stay alive?

List student responses on the board or on chart paper, then ask:

- If you placed a male and female mouse in a cage with food and water, what do you think would result? *(Depending on their knowledge, students may respond that mice reproduce frequently and the population would increase rapidly; that in certain cases, grown mice eat newborn mice; and that overcrowding may limit population growth.)*

Again, list student responses on the board or on chart paper. When there are no further responses you may wish to ask questions such as:

- How does this model show what happens to populations of organisms in nature? In what ways is it limited as a model?
- What would make this model more closely resemble the interactions of organisms in their environment?
- How is your ecocolumn similar to or different from this example?

"Unsupervised": Student Manual pages 36–38.

Divide the class into pairs. Have students read the Introduction and carry out the Procedure for "Unsupervised." This activity introduces the effects of limiting factors such as food, space, and waste accumulation on population size. As you circulate among the pairs, check to see that students have set up their charts logically and constructed their graphs accurately.

TEACHING STRATEGY

If you choose to have students do the actual investigation in which they monitor the rise and fall of a yeast culture rather than using the preexisting data provided in this activity, have each group prepare a 50-mL flask with 20 mL of a 10% molasses solution. Have them add a few drops of a yeast solution then swirl the flask. Have students place one drop of stirred solution from the flask on a clean microscope slide (preferably one with a grid) and, under high power (400X), count and record the number of yeast cells in the different fields of view. Leave the flask at room temperature, swirling it occasionally. Have students sample and count the culture each day until the yeast population has plateaued. Students will not see death unless they do a viable count.

▶ HOMEWORK

Have students complete the graphs and write responses to the Analysis that follows the activity "Unsupervised."

Session Two

DISCUSSION QUESTIONS Begin this session by having students discuss the Analysis that they did for homework. To facilitate the discussion you may wish to ask the following (modified from the Student Manual):

- During which time intervals was the population growth most rapid? Why?
- What happened during the other time intervals?
- What are the stages of growth on your graph?
- What if the yeast were grown in a larger flask? How would your graph look different? Why do you think so?
- How is this growth curve similar to and different from the exponential growth curve? What might be some reasons for the difference? *(Student responses should include that the J-curve developed through 72 hours, but then the organismal death rate outpaced the birth rate and the population size decreased. The lack of food or space to sustain the population and the buildup of waste could be some of the reasons for the difference.)*
- What is meant by "carrying capacity?" What might cause it to change? *(Student responses should include specific limiting factors.)*

Experimenting and Investigating

Divide the class into groups of four. Have students read the Introduction and follow the Procedure for the activity "The Lynx and the Hare." As students carry out the predator-prey simulation, circulate among the groups and determine if students are following the procedure carefully. You may wish to ask the following:

- What causes a lynx to die? to reproduce?
- What causes a hare to die? to reproduce?

ASSESSMENT

▷ Do students understand the factors that affect the growth or reduction of populations?

▷ Do students understand the relationships between population changes and the resultant J-curve or S-curve?

▷ Are students clear about what is happening at each section of the growth curve in Figure 4.3?

"The Lynx and the Hare": Student Manual pages 38–40.

See page xv in the Introduction section of this Teacher Guide for information on the use of **Models** in science classrooms.

Notes

▶ **HOMEWORK**

Have students write responses to the Analysis that follows "The Lynx and the Hare" activity.

If the snail/*Elodea* activity from Learning Experience 4 is concluded, remind students to write responses to the Analysis of "It's Elemental."

Processing for Meaning
Session Three

DISCUSSION QUESTIONS Begin class by having each group show the graph from "The Lynx and the Hare" activity. You may wish to facilitate discussion by asking:

- In what ways are groups' graphs similar? In what ways are they different? How do you account for this?

- How was the size of each population dependent on the other? *(Students should state that soon after the increase in hare population, there is an increase in the lynx population. Likewise, following the decrease in hare population, there is a decrease in the lynx population.)*

- What other factors could affect the size of the hare population? *(Students should mention factors such as the availability of food or shelter, climate, disease, old age, natural disaster, predation by other species, competition with other species, etc.)*

- What other factors could affect the size of the lynx population? *(Students should mention other prey or predators in the ecosystem, climate, disease, loss of habitat, natural disaster, hunting, etc.)*

- What would happen if all of the hares were removed? *(Students should respond that if the hares were the sole prey of the lynx in this ecosystem, the lynx population would dwindle and die out.)*

- What would happen if all the lynx were removed? *(Students should respond that the lynx were the sole predator, generally the hare population would undergo exponential growth. However, environmental factors such as a drought could reduce the amount of food available, causing the hare population to dwindle if it exceeds its carrying capacity.)*

ASSESSMENT

◉ Have students plotted their graphs correctly?

◉ Do students realize that their graph shows only the effect of predation on the two populations?

◉ Have students identified other factors that may influence changes in the lynx and hare populations?

SCIENCE BACKGROUND

The hare and lynx population fluctuation are thought to be more complex than this activity suggests. Ecologists have noticed that hare populations on islands where lynx are not present show similar cyclic fluctuations. Periodic crashes in the hare population may be associated with changes in the avail-

Learning Experience 4   Population Pressures

ability of their plant food supply. Most patterns of population growth are caused by multiple interacting factors rather than a single predator-prey relationship.

DISCUSSION QUESTIONS Continue class by having students discuss their results to the "It's Elemental" activity found in Learning Experience 3 with the following questions (also found in the Student Manual).

- In which tubes did you note a color change? What caused the change in each tube? *(Students should respond that bromothymol blue changed from blue to yellow [or green] in tubes 2 and 4.)*
- Which organisms in which tubes remained healthy? Why?
- Which gases were cycled in tubes 2-4? How do you know? *(Students should respond that carbon dioxide cycled from the snail to the* Elodea *and oxygen from the plant to the snail. Both organisms in tube 4 remained healthy.)*
- What do you think would happen in each tube if they were placed in the dark? *(Students should respond that organisms in all the tubes would die since the plant would not be able to carry on the process of photosynthesis in the dark.)*
- In what group do the test tubes resemble the ecocolumns?

Applying

During the remaining time, introduce students to the case study "A Friendly Warning." The reading describes the relationships among five different populations of organisms in a forest ecosystem. Explain to students that they are to analyze the information presented in the article, use their knowledge of the interactions of organisms in an ecosystem and of population dynamics, and predict what will happen in this ecosystem during the next 10 years.

Distribute the non-text illustration "Acorn Rollercoaster" to each student (see Advance Preparation). Suggest that they take notes on the illustration as they read. Have students read the Analysis that follows the case study. Review the main points of the Analysis and describe your expectations for their completed project.

> "A Friendly Warning": Student Manual pages 41-44.

SCIENCE BACKGROUND

Ticks undergo three stages of development during their incomplete metamorphosis. Tick eggs develop into larvae that hatch into nymphs, or small ticks. At each successive growth stage, the tick molts, or sheds its exoskeleton, until it is in its adult form.

▶ HOMEWORK

Have students read the article "Ticks and Moths, Not Just Oaks, Linked to Acorns" and start to design the poster.

▶ EMBEDDED ASSESSMENT

This project may be used as an embedded assessment to determine student ability to apply concepts from the module. Students will be required to present in a poster the results of their analysis, describe the biotic relationships among the organisms in the ecosystem, analyze the population fluctuations, predict the population growth and decline of the organisms for a 10-year period, and graph their predictions.

▶ SCORING GUIDE FOR EVALUATION OF STUDENT RESPONSES

SCORE	DESCRIPTION
Level 4	Student poster and response is clear and detailed and contains all the components listed below. Student shows the relationships among the five organisms, predicts future population fluctuations accurately, and relates Lyme disease and defoliation to the relevant population increases.
Level 3	Student poster and response is clear, detailed, and contains most of the components listed. Student shows the relationships among the organisms, predicts future population fluctuations, and relates occurrence of Lyme disease and defoliation to fluctuations.
Level 2	Student poster is less clear and detailed and contains some of the components listed. Student response shows partial understanding of the relationship of Lyme disease and defoliation to population fluctuations.
Level 1	Student poster is neither clear nor detailed and shows little understanding of the relationships among the populations of organisms, Lyme disease, and defoliation.
Level 0	Student poster is totally incorrect or nonexistent.

A complete poster with labeled graph includes a clear and accurate presentation of the following:

- biotic relationships showing
    - acorn bumper crop occurring every 2-4 years (except following defoliation by the Gypsy moth)
    - deer population increasing in the summer of a bumper crop
    - mice population peaking in the summer following a bumper crop
    - tick population increasing in the summer after a bumper crop (same time as mouse population increases)

EMBEDDED ASSESSMENT

- mice population decreasing sharply following a peak summer
- moth population building for several years following the mice population decrease
- moth population decreasing sharply following severe defoliation

• clear depiction of each organism's peak and fall
• numbers that are representative of naturally occurring populations
• analysis of prediction of population fluctuations showing:
  - peak years for Lyme disease follow tick population increase
  - peak years for defoliation follow moth population increase
  - defoliation causes a delay in bumper crop of acorns

*See page iv in the Introduction section of this Teacher Guide for suggestions on how to use **scoring rubrics for assessment**.*

In order to give time for students to create this poster clearly, accurately, and with some creativity, assign this project and then allow time for students to work on it at home. The processing for the case study will be found at the end of Learning Experience 5.

Session Four (to take place about a week after Session Three)

DISCUSSION QUESTIONS   Have students place their posters on the wall. Ask volunteers to describe their presentations and the conclusions they made. Encourage other members of the class to discuss the posters or ask questions. To continue the discussion, you may wish to ask questions such as:

◆ What patterns of interactions are evident in the graphs?

◆ Where have you seen these patterns before in the module?

◆ How does the forest's natural mechanism suggest ways for humans to interact with the environment?

◆ Barry Commoner, an ecologist, stated that "The first law of ecology is that everything is interconnected." Would this case study be a good example of the "interconnectedness" of living organisms? Explain your answer. Use specific examples.

ACORN ROLLERCOASTER

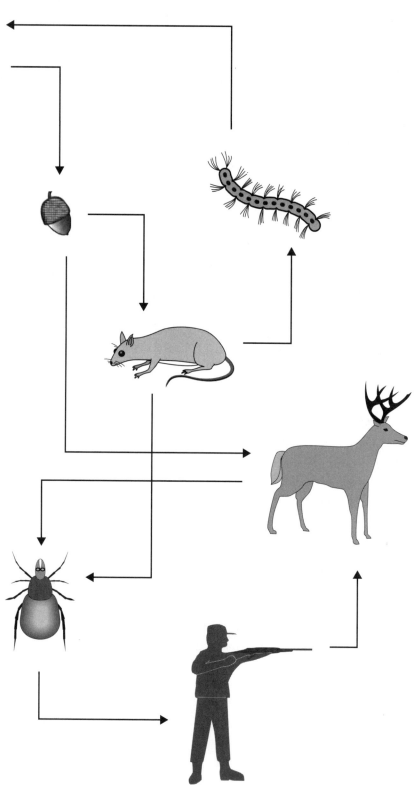

ACORN ROLLERCOASTER

Oak trees produce many acorns.

More acorns mean more mice.

Fewer gypsy moths mean healthier trees.

More mice mean fewer gypsy moths.

More deer come to the forest.

More deer mean more ticks.

More mice mean more ticks with Lyme spirochetes.

More ticks mean more Lyme disease.

More deer are available for hunters.

Learning Experience 4   Population Pressures

FOR FURTHER STUDY

BE FRUITFUL AND MULTIPLY?

OVERVIEW If human populations continue to grow at their present rate, how will that growth affect environmental quality, ecosystem on Earth, and human life? The world population has doubled since 1950, and the current trajectory may lead to dire consequences for our planet and all of its inhabitants. In this learning experience, students will create a graph showing human population levels throughout history. After examining this graph, students discuss factors that contribute to the continuing growth curve and think about what this information might mean when thinking about Earth's carrying capacity.

LEARNING OBJECTIVES

- Students trace and describe the dynamics of human population growth by interpreting graphs.

- Students identify major human and environmental factors affecting human population growth in different areas of the world.

▶ SUGGESTED TIME

- 2 class sessions (45–50 minute periods)

▶ MATERIALS NEEDED

For each student:
- 2 sheets of graph paper
- cellophane tape

For the class:
- 1 knife (optional)
- 1 apple (optional)

▶ ADVANCE PREPARATION

1. Photocopy the student pages which are provided as Blackline Masters at the end of this learning experience. Make one copy for each student.

▶ ASSUMPTIONS OF PRIOR KNOWLEDGE AND SKILLS

- Students are familiar with the resources necessary for maintaining life.
- Students are familiar with creating and interpreting graphs.

▶ TEACHING SEQUENCE PREVIEW

SETTING THE CONTEXT

- Students create and then analyze a graph showing human population levels through history.
- Students discuss factors that may be responsible for the dramatic rise in human population.

PROCESSING FOR MEANING

- Students read "Putting the Bite on Planet Earth" and explore the relationships among populations, industrialized/developing countries and resource use.

FOR FURTHER STUDY

TEACHING SEQUENCE

Setting the Context

Session One

Early in human history, the population grew very slowly. Technology and medicine have increased the human life span and decreased infant mortality to the point where the human population is doubling approximately every 40 years. In this Setting the Context activity, students determine that humans are now in an exponential or J-curve growth pattern. Based on their own experience and previous education, students speculate on human and environmental factors that may contribute to this type of growth and consider what the growth curve might mean in relationship to Earth's carrying capacity.

In the activity "Is More Better?" students create a graph of the history of human population growth from the year A.D. 1 to the present, and projected to A.D. 2050. Divide the class into pairs. Have each pair read the Prologue, "Is More Better?" and carry out the Task and discuss and record their responses to the Analysis that follows.

> Prologue and "Is More Better?":
> Teacher Guide pages 71-72.

DISCUSSION QUESTIONS Gather the class to discuss in more detail the factors that are related to the exponential growth curve in their graphs. To facilitate this discussion you may want to ask the following (also found in the Student Manual):

- How does this growth curve compare to the growth curves in Learning Experience 4—Population Pressures?
- When do you notice a sharp rise in the curve?
- What do you think caused such a rise? *(Depending on how well students understand history, responses might include such things as human longevity, diseases cured or brought under control, the Industrial Revolution, and changes in agricultural methods and productivity.)*
- Do you think the world population will ever reach 12 billion? What human and environmental factors do you think would be necessary to allow for the world population to reach that number?
- Do you think there is a population problem in the world? If yes, what potential solutions might you suggest? If no, at what point do you think there might be a problem? Why then?

ASSESSMENT

- Are students creating an accurate representation of the data?
- Are students analyzing the graph in order to respond to the Analysis that follows the activity?
- Are students applying what they know about the carrying capacity of other organisms to that of humans?

SCIENCE BACKGROUND

Until the Middle Ages, human populations were held in check by disease, famines, and wars that made life short and uncertain for most people. Among the most destructive of natural population controls were the

For Further Study   Be Fruitful and Multiply?

FOR FURTHER STUDY

bubonic plague and other diseases that periodically swept across Europe between 1348 and 1650. During the worst plague years, between 1348 and 1350, it is estimated that at least one-third of the European population perished.

Human population size began to increase rapidly after A.D. 1600. Many factors contributed to rapid growth. Increased sailing and navigation skills stimulated commerce between nations. Agricultural developments, better sources of power, better health care and hygiene (particularly clean drinking water), and sanitation have allowed the population size to rise steeply.

DISCUSSION QUESTIONS If time permits, continue the discussion by having students consider the many ways in which humans have "improved" their surroundings in order to increase both the quality of life and the number of people that the world can support. You may wish to ask:

◆ Think about some of the things you need and use. Where does each object come from? Where was it grown or manufactured? What was needed to do that?

◆ What are some inventions or changes during the past 50–100 years that have increased the length and quality of human life?

▶ HOMEWORK

"Putting the Bite on Planet Earth": Teacher Guide pages 73–76.

Have students read the case study "Putting the Bite on Planet Earth" and write responses to the Analysis that follows. This article introduces world demographics and looks at Earth's capacity to support human beings.

Processing for Meaning
Session Two

DISCUSSION QUESTIONS Begin this session by discussing student responses to the Analysis they did for homework. This discussion should highlight the idea that studying populations is not simply looking at numbers of births and deaths. Instead it looks at population growth in light of demographics and human use of natural resources. You may wish to facilitate the discussion by asking the following (modified from the Student Manual):

◆ What is the most populous region in the world? Which area is expected to show the greatest increase in world population?

◆ What percentage of the world population lives in the United States?

FOR FURTHER STUDY

- What percentage of the world's population consumes only 25% of the world's energy and contributes only 10% of the hazardous wastes? What does this mean?
- What observations can you make about the relationships among population, industrialized/developed nations, and resource consumption?
- Think about what you eat, what you wear, what you own, and other things you come in contact with, the places you go in a single day. What natural resources do you rely upon on a daily basis?
- Would you consider lowering your standard of living to make more resources available to your children and grandchildren? To people in developing countries? What would you choose to give up? Why?

TEACHING STRATEGY

You may wish to help students visualize the percentage of the earth's surface that is capable of supporting human life by doing the following demonstration: Display an apple and tell students to think of it as Earth. Slice the apple into quarters. One quarter will represent the total land area of the earth. What does the remaining represent? Slice the land quarter in half and set one half aside. This represents the inhospitable land (polar ice caps, rock, mountains, swamps, etc.) The fraction that is left (1/8) is human habitat. Slice the remaining 1/8 into four equal sections, and set aside 3 sections. Tell students that the three pieces set aside represent areas too rocky, wet and cold, or with soil too poor to produce food. This represents the cities, factories, parking lots, and other areas that people inhabit. Then carefully peel the remaining slice and explain that the peel represents the soil upon which humankind depends. Explain that there is this fixed amount of food-producing land and that even with advanced technology each person's portion becomes smaller when the population increases and the quality of air, land, and water decreases.

For the remainder of the session, you may wish to discuss one or more of the following open-ended questions, focusing on Earth's limited resources and its carrying capacity. Many of these questions ask students for their opinions and cover potentially controversial topics. As students respond to the questions and to each other's comments, have them identify their responses as personal opinions. Remind students to be respectful of others who may have differing opinions.

TEACHING STRATEGY

Pilot and field test teachers noted strong prejudicial attitudes when students discussed population dynamics. You may wish to plan for this reac-

FOR FURTHER STUDY

tion with questions that might help students evaluate human population growth without blaming those who live in developing countries, or those living in poverty in developed countries. As students discuss world-wide population demographics and resource use, they may bring up controversial issues such as birth control, abortion, women's rights, education, and science technologies.

ASSESSMENT

- Are students using concepts from the module in their responses?
- Are students points of view based on knowledge or on emotion?
- Are students listening thoughtfully to other students' comments?

♦ What responsibilities, if any, do developed and developing nations have in making resources available to one another? Who should have the rights to use a limited resource? Why?

♦ Do you think human population growth is outpacing Earth's ability to house and feed humans? If so, whose responsibility is it to feed the starving? Why?

♦ Which is more important: preserving the natural environment in which resources exist or human need to use the resources? Why?

♦ How are other organisms affected by the way in which humans use natural resources? What do you think about these effects?

FOR FURTHER STUDY

BE FRUITFUL AND MULTIPLY?

PROLOGUE   How does the human population growth curve compare to that of other organisms you have studied? Is Earth capable of supporting everyone and everything? What factors influence this growth? By mid-1992, the world population had passed 5.5 billion. About 92 million more people are added to the world each year. But what do these numbers mean?

As you have seen, there are many factors involved in studying the population dynamics of a single species. The addition of 92 million people each year cannot be studied out of context. What are some of the factors that regulate the human population? Are they the same factors that influence other organisms? In this learning experience, you will explore human population dynamics, what the growth curve looks like, and how the factors that influence human population growth. You will also begin to investigate the carrying capacity for humans on Earth.

IS MORE BETTER?

INTRODUCTION   Do you think there is a world population problem? A population grows for as long as the number of births exceeds the number of deaths. Not only do the babies born each year increase the population by their own number, but they will, in turn, further enlarge the population when they grow up to have children of their own. Think about this in your family: How many children are in your family? How many children do you and your siblings think you would like to have? Now, think about the rest of the class and the number of people in their families and the number of children they think they will have and so on for a few generations. If you were to think about this for the whole United States you would see a trend towards increasing population. However, when considering population dynamics, one must also factor in death rates.

FOR FURTHER STUDY

What can we learn by studying the human population growth curve? What is the projected size of the human population in the year 2025? In the following activity, you will graph human population levels through history. Based on your understanding of the growth curves from Learning Experience 4, you will then speculate on the factors that influence human population.

▶ MATERIALS NEEDED

For each student:
- 2 sheets of graph paper
- cellophane tape

▶ TASK

1. Join two sheets of graph paper.
2. Using the data in Table 4.1, make a graph of the world population from the year A.D. 1 to the present, and projected to A.D. 2050.

▶ ANALYSIS

Write responses to the following in your notebook.

1. How does this growth curve compare to the growth curves in Learning Experience 4—Population Pressures?
2. When do you notice a sharp rise in the curve?
3. What do you think caused such a rise?
4. Do you think the world population will ever reach 12 billion? What human and environmental factors do you think would be necessary to allow for the world population to reach that number?
5. Do you think there is a world population problem? If yes, what potential solutions might you suggest? If no, at what point do you think there might be a problem? Why then?

Table 4.1

From Understanding Our Environment, *by Ann Tweed, National Science Teachers Association, 1995, p. 30.*

YEAR (AD)	POPULATION (IN MILLIONS)
1	170
200	190
400	190
600	200
800	220
1000	265
1100	320
1200	360
1300	360
1400	350
1500	425
1600	545
1700	610
1750	760
1800	900
1850	1,210
1900	1,625
1950	2,515
2000*	6,228
2025*	8,472
2050*	9,000

*projected population

PUTTING THE BITE ON PLANET EARTH

(Excerpt from Putting the Bite on Planet Earth, Don Hinrichsen, International Wildlife, *September/October 1994, pp. 36–41.)*

Each year, about 90 million new people join the human race. This is roughly equivalent to adding three Canadas or another Mexico to the world annually, a rate of growth that will swell human numbers from today's [1994] 5.6 billion to about 8.5 billion by 2025.

These figures represent the fastest growth in human numbers ever recorded and raise many vital economic and environmental questions. Is our species reproducing so quickly that we are outpacing the Earth's ability to house and feed us? Is our demand for natural resources destroying the habitats that give us life? If 40 million acres of tropical forest—an area equivalent to twice the size of Austria—are being destroyed or grossly degraded every year, as satellite maps show, how will that affect us? If 27,000 species become extinct yearly because of human development, as some scientists believe, what will that mean for us? If nearly 2 billion people already lack adequate drinking water, a number likely to increase to 3.6 billion by the year 2000, how can all of us hope to survive?

The answers are hardly simple and go beyond simple demographics, since population works in conjunction with other factors to determine our total impact on resources. Modern technologies and improved efficiency in the use of resources can help to stretch the availability of limited resources. Consumption levels also exert considerable impact on our resource base. Population pressures work in conjunction with these other factors to determine, to a large extent, our total impact on resources.

For example, although everyone contributes to resource waste, the world's bottom-billion poorest and top-billion richest do most of the environmental damage. Poverty compels the world's 1.2 billion bottommost poor to misuse their environment and ravage resources, while lack of access to better technologies, credit, education, health care and family-planning condemns them to subsistence patterns that offer little chance for concern about their environment. This contrasts with the richest 1.3 billion, who exploit and consume disproportionate amounts of resources and generate disproportionate quantities of waste.

One example is energy consumption. Whereas the average Bangladeshi consumes commercial energy equivalent to three barrels of oil yearly, each American consumes an average of 55 barrels. Population growth in Bangladesh, one of the poorest nations, increased energy use there in 1990 by the equivalent of 8.7 million barrels, while

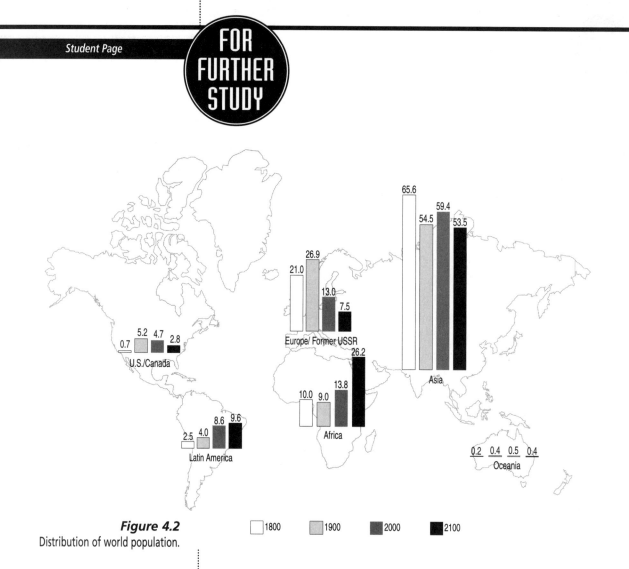

Figure 4.2
Distribution of world population.

U.S. population growth in the same year increased energy use by 110 million barrels. Of course, the U.S. population of 250 million is more than twice the size of the Bangladeshi population of 113 million, but even if the consumption figures are adjusted for the difference in size, the slower growing U.S. population still increases its energy consumption six or seven times faster yearly than does the more rapidly growing Bangladeshi population.

In the future, the effects of population growth on natural resources will vary locally because growth occurs unevenly across the globe. Over the course of the 1990s, the Third World's population is likely to balloon by more than 900 million, while the population of the developed world will add a mere 56 million. Asia, with 3.4 billion people today, will have 3.7 billion people by the turn of the century; Africa's population will increase from 700 million to 867 million; and Latin America's from 470 million to 538 million. By the year 2000, the Third World's total population is expected to be nearly 5 billion; only 1.3 billion people will reside in industrialized countries.

The United Nations estimates that world population will near 11.2 billion by 2100. However, this figure is based on the assumption that

growth rates will drop. If present rates continue, world population will stand at 10 billion by 2030 and 40 billion by 2110...

Perhaps the most ominous aspect of today's unprecedented growth is its persistence despite falling annual population growth rates everywhere except in parts of Africa, the Middle East and South Asia. Annual global population growth stands at 1.6 percent, down from 2 percent in the early 1970s. Similarly, the total fertility rate (the average number of children a woman is likely to have) has dropped from a global average of six only three decades ago to slightly more than three today.

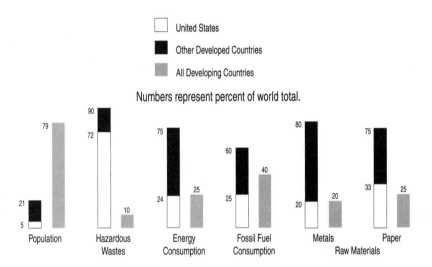

Figure 4.3
Populations and resource use.

Population continues to grow because of tremendous demographic momentum. China's annual growth rate, for example, is only 1.2 percent. However, the country's huge population base—1.2 billion people—translates this relatively small rate of growth into a net increase in China's population of around 15 million yearly. Clearly, any attempt to slow population growth is a decades-long process affected by advances in medicine, extended life spans and reduced infant, child and maternal mortality.

▶ **ANALYSIS**

Write responses to the following in your notebook.

1. Look at Figure 4.2. What is the most populous region in the world? Which area is expected to show the greatest increase in world population?

2. Look at Figure 4.3. What percentage of the world population lives in the United States?

3. What percentage of the world's population consumes only 25% of the world's energy and contributes only 10% of the hazardous wastes? What does this mean?

4. What observations can you make about the relationships among population, industrialized/developed nations, and resource consumption?

5. Think about what you eat, what you wear, what you own, and other things you come in contact with, the places you go in a single day. What natural resources do you rely upon on a daily basis?

6. Would you consider lowering your standard of living to make more resources available to your children and grandchildren? To people in developing countries? What would you choose to give up? Why?

Variation...Adaptation... Evolution

OVERVIEW In this module, students have been exploring the interaction among various organisms and their environments. Whether organisms succeed in these interactions depends on their structural and functional adaptations; adaptations that have evolved over time and led to the variety of life on Earth. But how did such variety of organisms come about and why did multitudes of other organisms become extinct?

The process of evolution, principally by natural selection, is the underlying tenet of biological science: It provides an explanation both for the unity and diversity of life and for the adaptation of organisms to their environment. Random changes in the DNA of an organism and the subsequent expression of these genes cause new characteristics to appear. While such mutations are often deleterious, sometimes they enable an organism to be better suited to its environment and better adapted to carry out functions needed for life. Changes in the gene pool of a species are the basis for all evolutionary changes and, ultimately, for the emergence of new species.

Humans have grouped living organisms into species as a means of comprehending and sorting out the incredible biological diversity around us. Classifying organisms by appearance, by function, and by specific structures has helped provide order to the study and identification of living things. But the identification of separate species is not always so straightforward; sometimes two species are quite clearly different by visual criteria, but in other cases the difference can be established only by elaborate techniques and measurements.

In this learning experience, students explore how the structures of organisms are correlated to their functions; learn how adaptations, as a form of evolutionary change, arise and are related to the environment; and determine that a group of organisms capable of interbreeding constitute a species.

LEARNING OBJECTIVES

- *Students determine how structure is related to function in objects and organisms.*

- *Students explain Darwin's theory of natural selection, and compare the theories of gradualism and punctuated equilibrium.*

- *Students explore the random nature of variations which, in conjunction with the environment, provide the basis for natural selection.*

- *Students examine the criteria used to identify groups of organisms as species.*

- *Students define a species as a group of organisms capable of interbreeding and producing fertile offspring, and examine the biological factors which determine whether two populations will interbreed.*

SUGGESTED TIME

- 4 class sessions (45–50 minute periods)

MATERIALS NEEDED

For each pair of students:
- 1 paper bag with strips (see Advance Preparation)
- 2 sheets of graph paper
- 2 pens or pencils (of different colors)
- label or index card
- "mystery" object such as
  - potato ricer
  - strawberry huller
  - olive pitter
  - dandelion digger
  - egg slicer
  - hose nozzle
  - nutcracker
  - protractor
  - binder clip
  - hardware item, etc.

ADVANCE PREPARATION

1. Prior to Session One, locate enough "mystery objects" to provide one for each pair of students. Number the mystery objects in consecutive order. Set up a "station" for each pair of students and place one at each object at each station.

2. You may wish to gather illustrations that show plant and animal structures for the discussion that follows the activity "What Is This, Anyway?"

3. Prior to Session Two, photocopy "Movement Directions" provided as a Blackline Master at the end of this learning experience. Make one copy for each pair of students, cut into strips, and place one set in each paper bag.

ASSUMPTIONS OF PRIOR KNOWLEDGE AND SKILLS

- Students understand that a gene is a segment of a DNA molecule that codes for a specific protein. This molecular transfer of information determines the characteristics of organisms.
- Students are familiar with concept-mapping.

▶ **TEACHING SEQUENCE PREVIEW**

SETTING THE CONTEXT

- Students predict the functions of objects by observing their structures in the activity "What Is This, Anyway?"
- Students read "Evolutionary Theory: Past and Present" and relate the structure-to-function linkage to Darwin's theory of natural selection.

EXPERIMENTING AND INVESTIGATING

- Students explore random changes in the activity "Going with the Flow."
- Students decide which changes and adaptations are selected as "fittest."

PROCESSING FOR MEANING

- Students discuss how random variations accumulate in a population, and that only variations allowing adaptation to an environment will be selected for in that environment.
- In the activity "Of Fish and Dogs," students identify criteria they would use to group organisms into species.
- Students determine that habitats, feeding habits, and the ability to reproduce are used to define species.

APPLYING

- Students discuss other factors that lead to speciation, such as chromosome number and geographical isolation.

Setting the Context

Session One

Why are species' structures matched to the functions they perform? What has caused this matching of structure to function? A deciduous leaf, for example, is broad and flat enabling it to attain maximum sunlight, has an outside cuticle that prevents water loss and a stoma on the under surface allowing for gas exchange, and contains many chloroplasts within its cells. These macroscopic and microscopic structures enable the leaf to carry on the process of photosynthesis, the food-making process upon which all life depends.

In this Setting the Context, students begin by attempting to identify the function of some unfamiliar human-made objects from its structure. As they proceed from one object to the next, students determine that the structure of each object offers clues to its function.

Begin class by having students read the Prologue and "What Is This, Anyway?" Have students number a sheet of paper according to the number of "stations" you have set up (one per pair of students). Assign a pair of students to each station (see Advance Preparation). Remind them to record the predicted function of the first mystery object on the sheet not at number one, but at the number that corresponds to the object number. Allow two minutes at each station and direct students to move when you give a signal. Be sure to leave enough time at the end of the class for the discussion.

> **Prologue and "What is This, Anyway?":**
> Student Manual pages 47–48.

TEACHING STRATEGY

You may wish to have a numbered station diagram on the board to show the direction in which students should move from station to station. There should be only one pair of students at each object at a time.

When students have finished, allow a few minutes for them to respond to the Analysis in their notebooks. Students will want to find out what each object is and what it does. Start at station one, and hold up the object. Ask students for their predictions as to its function, and their reasons for the prediction. Continue in this way until the functions of all the objects have been predicted.

DISCUSSION QUESTIONS Continue by discussing student responses to the following (also found in the Student Manual):

- What general principle or concept can you state after doing this activity? *(Students should respond by stating that the structure of an object is generally directly related to its function.)*

- What might be the importance of this principle in the plant and animal world? Explain by using several examples. *(Students might respond that in nature an organism must have suitable structures with which to carry out the characteristics of life. Examples include the eyes, beak, and claws of a hawk which enable it to catch food; its wings to move, etc. Plants have many roots and root hairs for maximum water uptake, leaves for food making, colorful flowers with nectar to attract insects, which facilitate reproduction, etc.)*

To illustrate the structure-to-function concept, you may wish to show pictures of animal and plant structures such as a lobster's claws and antennae; a frog's tongue, jumping hind legs, and balancing front legs; and a cactus's wide stems and spines. You may want to have students speculate about the fate of a lobster without claws, or a frog without hind legs.

DISCUSSION QUESTIONS
Continue the discussion with the following question:

- In the natural world of organisms, which do you think might come "first," the structure or the function? Explain your response. *(Student responses will vary, however, the relationship of structure to function is integral to the continued existence of organisms. All structures are dependent on functions, and there can be no function without a structure.)*

SCIENCE BACKGROUND

This last question is designed to be thought-provoking and is somewhat specious: there is no ordered or determined designing of a structure to fit a function in the natural world. There is often a misconception among students that "polar bears have heavy coats because they live where it is cold." Structures have evolved from many small variations, and those variations may enable the organism to be suited to the environment and survive. Evolution is not predetermined to a needed function or end. In fact, the fossil record is full of organisms that had exquisite structures but no longer exist. These important concepts must be clear to students in order for them to understand why it may appear that there has been a deliberate match between a structure and its function. This is developed further in the reading "Evolutionary Theory: Past and Present" and in the activity "Going with the Flow," both of which are in the Student Manual.

▶ HOMEWORK

Have students read "Evolutionary Theory: Past and Present" and respond to the Analysis that follow the reading. This Analysis is designed to encourage students to integrate the structure-to-function concept with evolutionary changes over time and the theory of natural selection.

"Evolutionary Theory: Past and Present":
Student Manual pages 49–53.

Learning Experience 5   Variation...Adaptation...Evolution

Experimenting and Investigating

Session Two

DISCUSSION QUESTIONS Begin this class session by having students discuss their responses to the Analysis that they did for homework. The following questions are modified from the Student Manual:

- How would you explain Darwin's theory of natural selection and relate it to the concept of evolutionary change over time?

SCIENCE BACKGROUND

The word "theory" is much more definitive in science than it is in ordinary speech. In science, a theory is a time-tested concept that makes useful and dependable predictions about the natural world and which is based on many observations and experiments. Those who fear that evolution is anti-religion often dismiss it by saying that "it's only a theory." In reality, evolutionary change in organisms over time is a fact. The most comprehensive and accepted explanation of how organisms evolve is the theory of natural selection.

- How might you compare gradualism and punctuated equilibrium? *(Students should respond that both explain how organisms have changed throughout the history of life on Earth. Gradualism follows Darwin's thinking that there has been slow and gradual change in the types of populations due to selective [environmental] pressures. Punctuated equilibrium suggests most species remain relatively unchanged for most of their existence and that changes in organisms occur during a relatively brief periods. For example, catastrophic climatic changes after long periods of stability cause mass extinctions, with new organisms filling the now-empty niches. Both processes are corroborated by the fossil record.)*

- How might punctuated equilibrium explain gaps in the fossil record? *(Students should respond that there is often no fossil evidence of the gradual stages as populations change over time.)*

SCIENCE BACKGROUND

Although most evolutionary changes have occurred over millions of years, some have occurred in a very short period of time. The classic example is the peppered moth, in which most of the population changed from dark to light (accompanied by changes which reduced air pollution in England) over a period of 50 or so years. In *The Beak of the Finch (A Story of Evolution in Our Time)* (New York: Knopf, 1994), Jonathan Weiner describes current research on the same finches that inspired Darwin's thinking (see

Figure 5.1). Peter and Rosemary Grant return to the Galapagos yearly where they observe, weigh, measure, and track the finches on the island of Daphne Major, a tip of a young volcano. The finches can not easily fly away; there are rapid changes in the environment, and thus food sources also change. The Grants note variations in the beaks of the finches and also in the DNA taken from blood samples. Accelerated environmental change appears to cause an acceleration of evolution. Thus, what had been seen only in the fossil record is now visible within a human lifetime.

◆ What type of food do you think each finch eats? Why do you think so?

Name	Vegetarian tree finch	Large insectivorous tree finch	Woodpecker finch	Cactus ground finch	Sharp-beaked ground finch	Large ground finch
Shape of bill	Parrotlike bill	Grasping bill	Uses cactus spines	Large crushing bill	Pointed crushing bill	Large crushing bill
Main food	Fruit	Insects	Insects	Cactus	Seeds	Seeds
Habitat	Trees	Trees	Trees	Ground	Ground	Ground

Figure 5.1
Beaks of Darwin's finches and the finches' foods.

◆ Explain the role of the environment in both evolution and extinction. *(Student responses might include that environmental pressures determine which random variations and subsequent adaptations allow organisms to survive, thus giving rise to a population with new characteristics. Climatic changes, natural disasters, and human intervention cause extinctions.)*

SCIENCE BACKGROUND

In 1977, Walter Alvarez and his colleagues at the University of California at Berkeley discovered large amounts of the metal iridium in certain Cretaceous rocks in Italy, and subsequently, in other far-flung places on Earth. Iridium is relatively rare in the earth's crust, but is abundant in meteorites. They hypothesized that at the end of the Cretaceous period, a giant asteroid, about 10 kilometers in diameter, crashed to the earth, releasing a cloud of debris that circled the earth for months and darkened the sky. As a result of this collision and darkness, photosynthesis ceased, food chains collapsed, the climate changed, and terrestrial and marine organisms began dying off. Although there was fossil evidence that the extinctions of organisms continued over tens of thousands of years, the consensus was that the impact of an asteroid of that size would play a major role in the mass extinctions that occurred about 65 million years ago, including the demise of the dinosaurs.

In 1990, a buried crater, now estimated to be 200 kilometers in diameter, was found in Chicxulub in the Yucatan peninsula of Mexico. A variety of dating tests have determined the age of the crater at 65 million years.

Learning Experience 5   Variation...Adaptation...Evolution

Another smaller crater, about 32 kilometers in diameter, has been found in Iowa and is also dated at 65 million years. The Alvarez theory, as it is known, seems to answer the question: "What caused the extinction of the dinosaur?"

DISCUSSION QUESTIONS

You may wish to continue the discussion by asking the following questions:

- Why are the structures of organisms so well adapted to their functions? Give several examples. *(Students should respond that structures that enable an organism to carry out its life processes are better able to survive. Over time, there is an accumulation of favorable variations and thus the population is well adapted. For example, frogs have long, webbed hind legs for jumping and swimming, eyes at the top of the head for seeing above water, a tongue attached in the front for quick food-getting movement, and a greenish skin color for camouflage.)*

- Examine Figure 5.5 in the Student Manual. What do you observe? What conclusions can you draw? *(Students might respond that large extinctions happen on a fairly regular basis, that geologic time may be measured/named by these extinctions, and that earlier organisms appear to be marine, and structurally simpler than later terrestrial ones.)*

SCIENCE BACKGROUND

The history of Earth and many of the life forms that lived on it may be "read" in the rocks on and near its surface. The geologic eras were identified and named in the early twentieth century; the names of the period subdivisions relate to the areas where most of the rock strata were studied. Darwin realized that any theory of gradual evolutionary change depended on Earth having a long history. Radioactive isotopic measurements have confirmed that Earth is close to 5 billion years of age. Equally important is the fact that the ages of the fossils found within these rocks have also been dated. The fossil record has many gaps, in that intermediate fossils definitively showing each change are often not present. It was long assumed that the intermediate fossils once existed, but had been destroyed or had not been found yet—that they had been washed into inaccessible areas, for instance.

In 1972, Niles Eldredge and Stephen Jay Gould proposed that the fossil record may actually be complete. They noted that a species might appear suddenly in the record and when it disappeared, even millions of years later, it was not much different from when it appeared. Another related but different species might appear and also persist without much change. This led the two scientists to propose that long periods of stability followed by environmental stress and extinctions (punctuated equilibrium) are the history of life on Earth.

How might new species appear so suddenly? Eldredge and Gould speculated that allopatric speciation might be the answer, that is, that speciation among isolated populations occurred rapidly filling niches within the geographical area of the old parent species.

TEACHING STRATEGY

You may wish to refer to the National Association of Biology Teachers guidelines on the teaching of evolution. The NABT statement includes:
- The diversity of life on earth is the outcome of evolution: an unsupervised, impersonal, unpredictable and natural process of temporal descent with genetic modification that is affected by natural selection, chance, historical contingencies and changing environments.
- Students can maintain their religious beliefs and learn the scientific foundations of evolution.
- Teachers should respect diverse beliefs, but contrasting science with religion, such as belief in creationism, is not a role of science. Science teachers can, and often do, hold devout religious beliefs, accept evolution as a valid scientific theory, and teach the theory's mechanisms and principles.
- Science and religion differ in significant ways that make it inappropri-ate to teach any of the different religious beliefs in the science classroom.
(Excerpted from the "NABT Statement on Teaching Evolution," The American Biology Teacher, January 1996, adopted by the Board of Directors, March 15, 1995.)

The simulation "Going with the Flow" reinforces the concept of random variations as the source of natural selection. The variations in the imaginary creatures (signified by the commands) do not follow a plan or occur for a purpose. The organisms with variations that allow them to be well adapted to their environment survive and reproduce. The others die out.

Distribute a sheet of graph paper to each pair of students and have them fill in the center square of the grid with a pencil. Have students read the Introduction and carry out procedure steps 1–5 of the activity "Going with the Flow." Circulate among the pairs and listen to see if students understand that they are simulating random events in evolution.

"Going with the Flow": Student Manual pages 53-55.

MODULE CONNECTION

The *Insights in Biology* module *Traits and Fates* develops in greater detail the relationship between evolutionary theory and genetics. It is the random changes in the DNA that drive evolution; in effect, evolution is the change in a population's gene pool over time.

When students have finished Procedure steps 1–5, announce that you are the agent of natural selection and that only certain variations and adaptations (grid patterns) are suitable or most "fit" and will survive. The patterns might show: large, amorphous, diagonal chains;

compact, rectangular pattern with few gaps; large design with many gaps; etc. As the selective pressure agent, choose two or three shapes that are of the same pattern type. The pair patterns that match the designated shapes are the ones that are suitable, i.e., they are "selected" organisms and leave descendants. The others die out due to natural selection pressures.

TEACHING STRATEGY

It is best if you do not tell students the shapes that will be selected in advance. Do not choose recognizable shapes as this may reinforce student misconception that variations, adaptations, and thus, evolution, are directed and planned.

Have each pair of students whose grid has "died out" join another still-existing pair and have them follow Procedure step 7. Repeat the exercise again, selecting the same or different shape choices as before. Rearrange the groups of four as needed and continue for as many generations (rounds) as time permits.

▶ **HOMEWORK**

Have students write responses to the Analysis that follows "Going with the Flow" in their notebooks.

Processing for Meaning
Session Three

Begin this session by having students discuss the activity. The purpose of this discussion is for students to explain that each strip's direction is comparable to a slight random variation. The favorable variations accumulate in a population, but if these organisms are not adapted to the environment, they do not survive. Other organisms with more suitable adaptations survive, reproduce, and evolve over time.

DISCUSSION QUESTIONS Have students respond to the following (modified from the Student Manual):

♦ What does the central square of the grid signify? *(Students should respond that it is the starting point of the organism at this time.)*

♦ What were you showing as you followed the directions on the strips?

♦ Why did some grids not survive? *(Students should respond that those variations were not suitable and environmental pressures selected against their survival.)*

ASSESSMENT

▶ Do students understand that the "Movement Directions" strips represent random mutations?

▶ Do students understand that the environment "selects" the creature shapes that are best suited to survive?

- Why were the completed grids different from each other? *(Students should respond that the directions, or variations, were random and different.)*

- How can you explain your results by using Darwin's theory of natural selection? *(Students should explain that there are random changes [mutations] to the DNA and these cause variations in the organism. Those variations that are favorable allow the organism to adapt to its environment and survive. If many of these variations accumulate in many organisms in a population resulting in different structures or physiological functions, evolution has occurred.)*

- What do you think it would mean if another teacher came in and decided on different shapes? *(Students should respond that the environment would have changed and other random variations and adaptations would be more "fit" in this new environment.)*

- How might this activity illustrate the principle that individuals vary but do not evolve, that populations evolve? *(Students may have difficulty with this question. Each organism is born with its own variations. The grid pattern is the accumulated variations over generations and the pattern that "survives" is the one that is best adapted to its environment. Evolution occurs in the population over time.)*

Have students read the Introduction and Task for "Of Fish and Dogs" and work in pairs to examine Figures 5.7 and 5.8. Allow students about 10 minutes to complete the Task, then gather the class together and have each pair present their ideas about what constitutes a species and what criteria they would use to group organisms in species.

Keep a list on the board or on chart paper as students present their criteria. After each pair has had an opportunity to explain its decision and criteria, inform students that while each of the fish pictured in the illustration represents a different species of cichlids (*Haplochromas*) which are all found in Lake Victoria in Africa, all the dogs in the illustration are members of the same species, *Canis familiaris*. The differences observed in the dogs are characteristics of different breeds (groups), not different species.

DISCUSSION QUESTIONS Have students discuss why the fish, which look so similar, may be grouped in different species, while the dogs, which look so different, are all the same species. The discussion should focus on criteria that might be used to define a species, such as habitats, feeding habits, and, ultimately, the ability to reproduce. You may wish to facilitate discussion by asking:

- Where do dogs commonly live? Are the habitats determined by their type? That is, do cocker spaniels live in one kind of habitat, terriers in another, and Great Danes in yet another? Why or why not?

ASSESSMENT

▸ Have students correlated the concept that adaptations are favorable only if they permit a population to survive in its environment?

▸ Do students understand that even though it seems that organisms have been designed to survive in their environments, it is actually the results of selection pressures on random changes over long periods of time?

"Of Fish and Dogs": Student Manual pages 55–57.

- What do dogs normally eat? How is their diet reflected in the structure of some of their parts?

- Can different types of dogs mate and produce viable (living) offspring which are also capable of mating? Why or why not?

- Examine the illustrations of the cichlids carefully. Do you see differences among the structures of the different fish? *(Students may observe that the structure of the mouths are slightly different. This is very subtle in some of the fish and you may have to point it out.)*

- What might differences in mouth structure suggest to you about the habits of the fish? *(Students should make the connection that mouth structure is indicative of diet.)*

- What might different fish eat? *(Students responses should include a range of diets; insects, algae, plankton, other fish, plants, mollusks.)*

- How might their diets influence where in Lake Victoria these fish live? *(Student responses should indicate that diet is related to habitat; for example a diet of flying insects or algae require the fish to live close to the surface of the water whereas a diet of bottom-dwelling fish would have the cichlids living deeper in the lake.)*

- How might where fish live in the lake influence the mating habits of cichlids? How does this differ from the mating habits of dogs?

- What other factors might influence which fish mates with which fish? *(Students might suggest the importance of coloration and courting behavior on mating habits.)*

SCIENCE BACKGROUND

Cichlid diversity and specialization principally involves differences in cranial morphology and anatomy, dental morphology and body features relating to the trophic niches they occupy. Each species has specifically adapted dentition and jaw structure for its specific feeding habits. These reflect the specific diet of each species and include structures for feeding on boring insect larvae and pupae, feeding on arthropods, scraping algae, grazing on organismal cover on rocks, taking scales off other fish by biting and scraping, feeding on detritus, eating mollusks, preying on other fish (including embryos and newly shed eggs of other cichlids), feeding on ectoparasites of other fish and on zooplankton and phytoplankton and plants. It is these anatomical features which in part both determine the accessibility of different species to each other for reproductive purposes (feeding habits may isolate populations in different trophic niches in the lake) and provide recognition features for mating purposes; that is, fish with mouth structures suited for algae scraping may preferentially mate with other fish with the same mouth structures. Other components of mating recognition include species-specific coloration, pheromones, and courtship behavior patterns.

▶ **HOMEWORK**

Have students read "The Fundamental Unit" in which a definition of species is constructed, and write responses to the Analysis.

> "The Fundamental Unit": Student Manual pages 57–59.

Applying
Session Four

DISCUSSION QUESTIONS Begin this session by discussing student responses to the homework. The intention of this discussion is to help students construct a clear idea of criteria used to identify species and to begin to think about what forces in nature may have led to the appearance of so many different kinds of organisms. You may wish to begin this discussion with the questions that students addressed in their homework (modified from the Student Manual):

◆ Explain why the cichlids are considered different species and dogs the same species. Be sure to include in your explanation the factors which would or would not influence hybridization. *(Student responses should include the ideas that dogs inhabit the same habitats, eat the same kinds of food, are structurally very similar, and appear to mate freely and produce prodigious numbers of puppies among the different breeds. Thus they are considered one species despite their varied appearances. Cichlids, however, have very different habitats based on their feeding habits, and so different feeders may never encounter each other for mating. Even if they did they might not mate because of variation in mouth structure, coloration, pheromones, or courting behavior. So even though they look very similar they are different species).*

◆ How do you think the concept of finch species is related to the concept of cichlid species?

◆ Explain the mating of a horse and a donkey in terms of species. How does this extend your understanding of the definition of species? *(Students responses should indicate an understanding that despite being different species horses and donkeys can breed under artificial circumstances; this crossbreeding even overcomes the physical difference of different chromosome number but results in offspring that cannot reproduce themselves. So, it extends the understanding of species from Wilson's emphasis on the ability to interbreed to the larger definition of being able to produce fertile offspring.)*

Allow 10–15 minutes at the end of this session for each student to create an "evolution" concept map. Encourage students to use the readings in this learning experience as they create their maps. (See Figure 5.2 for an example of an evolution concept map.)

> See page xii in the Introduction section of this Teacher Guide for suggestions on teaching and using Concept-Mapping.

Learning Experience 5 Variation...Adaptation...Evolution

TEACHING STRATEGY

You may wish to collect and review these maps before going on to Learning Experience 6, in which the concepts of evolution and species will be expanded in the discussion of speciation and its role in the diversity of life forms found on Earth.

▶ HOMEWORK

Have students read "Cichlids Past and Present" and be prepared to discuss it at the beginning of Learning Experience 6.

Inform students that the posters for the cast study "A Friendly Warning" are due and that they should be prepared to present their conclusions to the class.

"Cichlids Past and Present":
Student Manual pages 59–60.

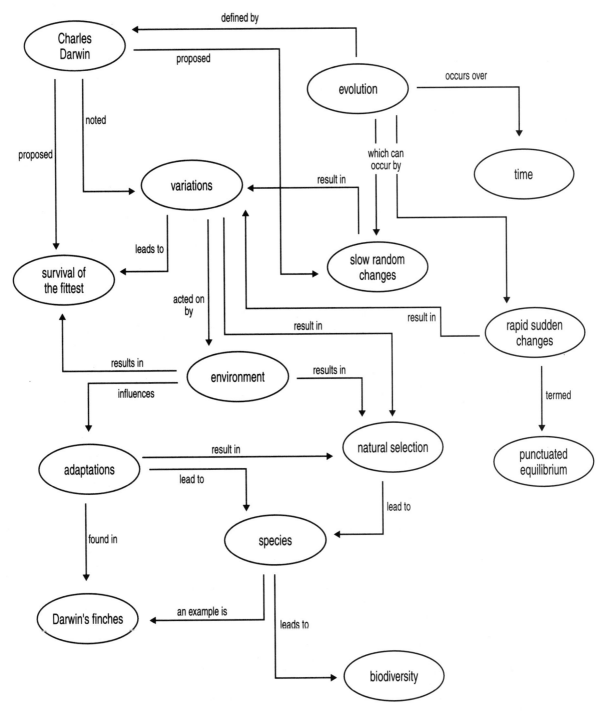

Figure 5.2
Evolution concept map.

MOVEMENT DIRECTIONS

ONE LEFT	ONE RIGHT
TWO LEFT	TWO RIGHT
ONE UP	ONE DOWN
TWO UP	TWO DOWN
REPEAT LAST STEP	REVERSE LAST STEP
NO CHANGE	NO CHANGE
DIAGONAL LEFT-UP	DIAGONAL RIGHT-UP
DIAGONAL LEFT-DOWN	DIAGONAL LEFT-UP

The Diversity of Life

OVERVIEW How do new species of organisms arise? By what mechanisms have species changed in the 3.8 billion years that life has existed on Earth?

In the last learning experience, students explored how the structures of organisms are correlated to their functions and how adaptations, as a form of evolutionary change, arise and are related to the environment. In this learning experience, they extend these ideas to see how new species arise (speciation). Students relate all these concepts to Darwin's theory of natural selection and finally to biodiversity.

Students then examine maps that illustrate patterns of species diversity (richness) in North and Central America and determine the reasons the patterns occur as they do.

▶ SUGGESTED TIME

- 3 class sessions (45–50 minute periods)

▶ MATERIALS NEEDED

For each group of three students:
- 1 set of "Species Richness" maps
- assorted colored pencils

For the class:
- reference material including atlases, encyclopedias, and geography books
- topographic, climatic, and landform maps of North and Central America
- 2 or 3 transparency sheets (optional)

▶ ADVANCE PREPARATION

1. You may wish to prepare transparencies of two or three student concept maps (homework for Learning Experience 5) that you feel diagram evolution well for use in Session One.

LEARNING OBJECTIVES

- ▷ Students examine examples of vertical evolution and divergent speciation.

- ▷ Students determine that biodiversity results from speciation, which results from geographic and reproductive isolation.

- ▷ Students compare patterns of mammal, bird, and tree diversity and speculate what the reasons for these patterns might be.

- ▷ Students correlate these patterns to the geography, latitude, and climate of the different regions.

2. Prior to Session One, prepare a transparency of the evolution concept map (see Figure 5.2).

3. Prior to Session Two, photocopy the three "Species Richness" maps found as Blackline Masters on pages 102–104 of this Teacher Guide, one set for each group of three students.

4. Read the entire learning experience including the *For Further Study* "Going…Going…Gone!" found at the end of this learning experience. If you choose to have the students conduct the *For Further Study* activity, read through the Advance preparation and order materials if necessary.

> **FOR FURTHER STUDY** In the *For Further Study* "Going…Going…Gone!" students explore the role of extinction in evolution. Students then examine data that suggests species are disappearing at an accelerated rate due to human activities. Finally, students think about the protection of certain endangered organisms and assess their own opinions regarding whether it matters if particular organisms become extinct.

See page viii in the Introduction section of this Teacher Guide for information and rationale of *For Further Study*.

▶ TEACHING SEQUENCE PREVIEW

SETTING THE CONTEXT

- Students discuss their concept maps on evolution.
- Students read "New Species" from E.O. Wilson's *The Diversity of Life* which describes evolutionary forces that lead to speciation.

EXPERIMENTING AND INVESTIGATING

- In "Mapping the Gradient across the Americas," students analyze patterns of species richness in North and Central America.

PROCESSING FOR MEANING

- Students discuss the factors that influence species richness patterns.

Setting the Context

Session One

Begin this session by posting, or showing transparencies of, several examples of the concept maps that students created in the last session. Engage students in a discussion of how the concepts are linked. Introduce the transparency of Figure 5.2 and use it to review the interconnections between Darwin's theory and other concepts that students explored in the last learning experience. This discussion should reinforce the concept that variations occur naturally, are passed on to future generations, and that those suitable to the environment become adaptations and survive. Evolution results from the total of these adaptations and lead to new species.

DISCUSSION QUESTIONS Have students read the Prologue and examine Figure 6.1 in preparation for a class discussion on the biodiversity of life. To facilitate the discussion, you may wish to ask the following:

> Prologue: Student Manual page 63–64.

- What does the phrase "the diversity of life" or the term "biodiversity" mean to you?
- Why is the elephant in Figure 6.1 so small? Why is the insect so large?
- What surprises you about this illustration? Why?
- What principle or concept is this illustration demonstrating? *(Students should respond that the pyramid shows those groups with the most species, i.e., the insects, are on the bottom and those with the fewest species, i.e., the mammals, are at the top.)*

SCIENCE BACKGROUND

The following is the key to Figure 6.1 in the Student Manual:

1. mammals
2. amphibians
3. bacteria
4. sponges
5. echinoderms
6. reptiles
7. coelenterates
8. birds
9. earthworms
10. roundworms
11. flatworms
12. fish
13. algae
14. protozoa
15. fungi
16. mollusks
17. noninsect arthropods
18. plants
19. insects

SCIENCE BACKGROUND

In the eighteenth century, a Swedish scholar named Carolus Linnaeus developed a system of classification which became the basis of all biological classification systems developed since then. He divided all plants into classes based on the number of stamens in each flower. Each class was then further divided into orders according to the number of pistils in the flower. He then divided each order into genera, and each genus into species, based on other structural differences. Linnaeus divided the animal kingdom into similar groupings, based on their physical characteristics.

Linnaeus' great contribution to classification was binomial nomenclature (two-term naming), his method of naming organisms by genus and species. These names were in Latin, a common language among scholars in all Western countries at that time. Dogs, for example, are classified as *Canis familiaris*. *Canis* is the genus which includes not only dogs but jackals, coyotes, and wolves, animals that they closely resemble structurally. But only domestic dogs are of the species *familiaris*.

Linnaeus' system provided the foundation for the modern-day system of classification which is more complex and, at times, controversial as scientists continue to attempt to identify criteria for defining relationships among organisms.

DISCUSSION QUESTIONS Continue the discussion with the following questions which are based on the reading "Cichlids Past and Present."

- How might you explain the evolution of a single ancestral species into over 300 species?
- What are some examples of human intervention in natural ecosystems?
- Were the results good, bad, or both? What are your reasons for saying so?

▶ **HOMEWORK**

"New Species": Student Manual pages 64–67.

Have students read "New Species" by E.O. Wilson and write responses to the Analysis. This reading discusses how vertical and divergent evolutionary forces lead to speciation.

Session Two

DISCUSSION QUESTIONS Begin class by asking students for their definitions of speciation. You may wish to write several of these definitions on the board. Have several volunteers write their flow charts on the board and explain their thinking. Continue the discussion with the following (modified from the Student Manual):

- How do geographic isolation and reproductive isolation lead to speciation? *(Students should respond that when organisms within a population are isolated, either geographically or reproductively, variations are selected by the new environment and new populations arise. Organisms from each of these populations are no longer capable of reproducing with each other or of producing fertile offspring, and thus new species arise.)*

Learning Experience 6 The Diversity of Life

- How are Darwin's finches an example of isolation and speciation? *(Students might speculate that an ancestral finch [that is, a pair or a female with a fertilized egg] somehow arrived from South America, giving rise to a population that became geographically and reproductively isolated from finches on the other islands. Students should note that certain variations of the finches survived at the expense of others, for example a variation in the beak made it better able to obtain food. The islands now have separate species of finches.)*

SCIENCE BACKGROUND

The Galapagos archipelago consists of 14 main islands created by volcanoes more than a million years ago (a young group in geologic time). The islands offer diverse habitats: dense thickets of thornbush and tree cactus in the lower elevations; tall tree ferns, orchids, lichens, and mosses in the more humid higher elevations.

Six species of ground finches inhabit most of the islands. They differ from each other mostly in the size and shape of their beaks (depending on their food supply). There are also six species of tree finches (four of which are insect-eaters), differing from each other in the size and shape of their beaks. The thirteenth finch looks more like a warbler; however, its internal anatomy classifies it among the finches. Although 12 of the finches look alike from the rear, males recognize females of their species by their beaks, and will not court females of another species.

The ancestral finch is thought to have been a small ground finch with a short, stout, cone-shaped beak adapted for crushing seeds. It is thought that founder groups of finches colonized the separate islands; however, once a group was established they became more geographically isolated with no gene flow among the different populations. (Finches are poor long-range fliers). Over time, 14 species of finches evolved. (Darwin collected thirteen. The fourteenth finch, a sharp-beaked ground finch, is extinct.)

- Do you think speciation is evolution? Explain your response. *(Students should respond that speciation is evolution, because variation and adaptation cause change in organisms over time and an accumulation of such changes results in new species [which increases biodiversity].)*
- What is the meaning of the quotation "The environment is the theatre and evolution is the play"? *(Students should respond that evolution takes place in, and is influenced by, the environment.)*

ASSESSMENT

▷ **Have students used appropriate terms and made logical connections?**

▷ **Do students understand that variations appear first and that selective environmental pressures determine whether they become adaptations?**

Experimenting and Investigating

The activity "Mapping the Gradient across the Americas" has students analyze patterns of mammal, bird, and tree diversity in North America and Central America and, using atlases and other geographical references, speculate as to the reasons for these patterns.

SCIENCE BACKGROUND

Biodiversity refers to the variety of life forms (plants, animals, fungi, and microorganisms) found on Earth, their genetic material, and their ecosystems. It is generally described as these three different components.

The first component, species diversity, is the variety of species found in an area. It can be measured in various ways. The three most common measures are species richness (the actual number of different species found in an area); species abundance (the number of individual members of each species present); and taxonomic diversity, which analyzes diversity based on the relatedness of the species in an area (the less related the species, the more diverse the area).

The second component, genetic diversity, is the amount of genetic variation within a species. The third component, ecosystem diversity, encompasses both the many differences in ecosystem type and the diversity of habitat and ecological processes within ecosystems, such as tropical rain forests, temperate forests, tundra, coral reefs, grasslands.

"Mapping the Gradient...": Student Manual pages 67–68.

Distribute one set of "Species Richness" maps (see Advance Preparation) to each group of three students. Instruct students to read the Introduction to "Mapping the Gradient Across the Americas," follow the Procedure, and write responses to the Analysis.

▶ HOMEWORK

In 1997, the National Aeronautical and Space Administration (NASA) landed the Pathfinder spacecraft on Mars, with the rover Sojourner on board to test the chemical composition of its rocks and soil. Introduce this homework by having students imagine that they are explorers from the planet Mars: Martian naturalists who think like a modern-day Darwin or Wilson and who now land in a rover in the students' own hometown. Have each student write an essay on what the Martians would see in this environment, using as much detail as possible. They should also include explanations (using concepts from the module) for what he or she sees. This exercise is designed to have students look at their surroundings from the perspective of an ecologist.

Processing for Meaning
Session Three

Begin the session by having volunteers read their Martian essays. You may wish to record student observations and to elicit further ideas from the class.

DISCUSSION QUESTIONS

Continue with a discussion of the "Mapping the Gradient Across the Americas" Analysis by asking (modified from the Student Manual):

- According to each map, which areas have the greatest number of species? The fewest?
- Which of the above areas are the same? Which are different? What might be some reasons for this?
- What patterns do you see that might explain the reasons for geographical species diversity? (*Students should respond that climate, especially rainfall and temperature, latitude [or geographical gradient], and proximity to oceans impact geographic diversity.*)
- What are some general principles that explain how geographical diversity reflects species diversity? (*Students should respond that there is an increase in diversity as one moves toward the equator. Climatic factors, especially rain and warmer air correlate to an increase in biodiversity. An area that is rich is one type of species is, generally, rich in other types of species.*)

ASSESSMENT

- Have students seen the pattern of increased biodiversity from the arctic toward the equator?
- Have students presented logical reasons for this geographical gradient?
- Have students noted inconsistencies and related them to climate and topography?
- Have students made the connection between evolution, speciation, and biodiversity?

SCIENCE BACKGROUND

The diversity of bird species increases by almost a factor of 10 from arctic areas to tropical areas. Mammals, too, show this distribution pattern. Tree species diversity for North America follows a similar pattern: The number of tree species again increases by a factor of almost 10 from the arctic toward the tropics. Mountain areas have much lower diversity than areas toward the coast. Similarly, desert areas in the Southwest are relatively poor in tree species when compared with temperate and coastal regions. Rain and sunlight encourage vegetal growth, as in the Southeast, and thus there is a greater number of species. In general, tree species diversity increases as the climate becomes warmer and more humid.

The geographical variation (latitude) observed in bird and mammalian diversity mirrors the underlying variation in tree species diversity. Areas that are rich in one form of biodiversity are likely to be rich in other forms as well.

Current ideas of why diversity increases as latitude approaches the equator include the following: the history of the earth (evolution and speciation), complex habitat (more niches), climate and climate variability (favorable conditions encourage more species, and species productivity.

Have students speculate as to possible reasons why there is a richness of species diversity near the equator.

- What has happened to the forests of Costa Rica? What might be some reasons? What might be the consequences?

Learning Experience 6 The Diversity of Life

▶ HOMEWORK

Have students write a comprehensive essay using as many of the concepts of this module as are applicable to the following statement: "The theories of evolution and speciation help to explain why there is such a diversity on life on Earth."

You may wish to write the statement on the board and have students copy it into their notebooks. Inform them that the essay should be at least 2–3 typed pages.

▶ EMBEDDED ASSESSMENT

The essay may be used as an embedded assessment. The following rubric is provided to help you assess student ability to understand the principles of evolution and speciation and to apply concepts from the module.

▶ SCORING GUIDE FOR EVALUATION OF STUDENT RESPONSES

SCORE	DESCRIPTION
Level 4	Student response is clear and detailed and contains all the components listed on page 101. Student shows an understanding of Darwin's theory of natural selection, relates variation and adaptation within populations to speciation, and explains how the diversity of life both influenced Darwin's thinking and results from speciation (vertical and divergent evolution). Student adds concepts from the module including those of habitat/niche, predator/prey, and others.
Level 3	Student response is clear and detailed and contains most of the components listed below. Student understands Darwin's theory of natural selection and how it relates to speciation and the diversity of life. Student adds few other concepts from the module.
Level 2	Student response is less clear and detailed and contains some of the components listed below. Student shows partial understanding of the theories of evolution and speciation and their relationship to the diversity of life. Student does not add other concepts from the module.
Level 1	Student response is neither clear nor detailed and show little understanding of evolution nor speciation nor their relationship to the diversity of life.
Level 0	Student response is totally incorrect or nonexistent.

EMBEDDED ASSESSMENT

A complete response includes clear and accurate presentation of the following four categories.

- Darwin's principles including:
  - organisms producing others like themselves
  - an overproduction of offspring
  - variation among offspring
  - variations that are passed on ("survival of the fittest") which are suitable to the environment
  - favored variations being adaptations ("selected" by the environment)
  - long periods of time being a contributory factor
- speciation is the evolution of new species including:
  - disappearance of the original species
  - influence of environmental factors
  - divergent evolution giving rise to new species over time
  - influence of geographical and reproductive isolation
  - influence of the habitat as to which species survive
  - new species evolving over time, not interbreeding
  - example of Darwin's finches on the Galapagos
- diversity of life on Earth including:
  - observed by Darwin on the *Beagle* voyage
  - speciation (a form of evolution) leading to diversity
  - influence of number of habitats and niches
  - extinctions leading to further diversity
- other concepts including:
  - evolution of habitats and their niches over the life of the earth
  - evolution of structures in organisms over time
  - coevolution of organismal relationships (predator/prey)
  - population fluctuations owing to biotic and abiotic factors

See page iv in the Introduction section of this Teacher Guide for suggestions on how to use **scoring rubrics for assessment.**

SPECIES RICHNESS—MAMMALS

Learning Experience 6   The Diversity of Life

SPECIES RICHNESS—TREES

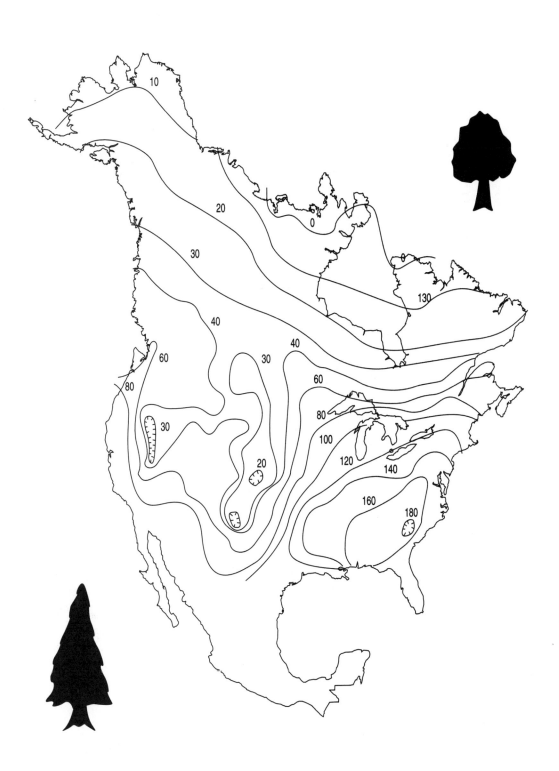

SPECIES RICHNESS—BIRDS

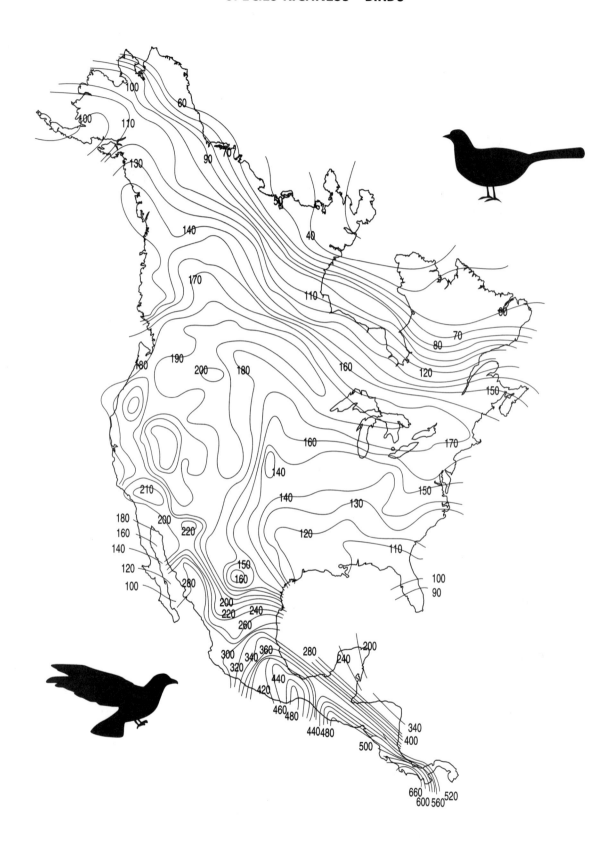

FOR FURTHER STUDY

Going...Going... Gone!

OVERVIEW Of all the species that have lived since life began on Earth almost four billion years ago, at least 99% are now extinct. This rate of extinction has played a central role in the evolution of species diversity. As species become extinct new niches become available and new, better adapted species appear to replace the former ones. Thus, extinction has played an important role in increasing biodiversity. And yet, for many the word extinction has taken on pejorative overtones. Today, the possible extinction of certain species of animals and plants is greeted with consternation and concern. Large amounts of money and political effort are spent on attempts to preserve many of our endangered species.

Does it really matter whether species are endangered and eventually lost? In this learning experience, students examine data that suggests species are disappearing at a rate greater than would be predicted from normal evolutionary causes and identify possible reasons for this accelerated extinction rate. Finally, after reading a case study on the Delhi Sands fly, students decide whether species extinction really matters.

▶ SUGGESTED TIME

- 2 class sessions (45–50 minute periods)

▶ ADVANCE PREPARATION

1. Photocopy the student pages which are provided as Blackline Masters at the end of this learning experience. Make one copy for each student.

LEARNING OBJECTIVES

▷ Students determine the factors that cause a species to become endangered and eventually extinct.

▷ Students determine the evolutionary and ecological consequences of extinction.

▷ Students decide whether it matters if a particular species becomes extinct.

▶ ASSUMPTIONS OF PRIOR KNOWLEDGE AND SKILLS

- Students are able to interpret graphs and charts.
- Students are familiar with concept-mapping.

▶ TEACHING SEQUENCE PREVIEW

SETTING THE CONTEXT

- Students discuss the concept of species extinction and interpret a graph of human population growth in comparison with a graph of the rising rate of extinctions over time.
- Students speculate about the factors that cause species to become endangered and/or extinct.

PROCESSING FOR MEANING

- Students discuss different types of extinction and the human behaviors that often lead to extinctions.
- Students read "What Good Is It, Anyway?" and create a concept map on the rainforest, its usefulness to humans, and the consequences that may result from its destruction.

APPLYING

- In the case study "Save the *Fly*? Are You Kidding?" students think about the idea of protecting certain organisms and not others and write an essay explaining their opinion.

FOR FURTHER STUDY

TEACHING SEQUENCE

Setting the Context

Session One

DISCUSSION QUESTIONS In this Setting the Context, students discuss the concept of species endangerment and extinction—its definition, its role in evolution, how it happens, and its effect on ecological systems. Have students describe both what they know about the concept of extinction and their views of it. You may wish to facilitate discussion with questions such as:

- What do the terms "endangered species" and "extinction" mean to you?

- Do you think that extinction is a positive process or a negative process? Explain your reasons.

- What kinds of things do you think cause extinction?

- Have you known of people who have been involved in or contributed time or money for projects designed to save an endangered species? If so, why do you think they felt it was important to do this?

- What factors do you think should determine whether a certain species is worthy of being "saved"?

Processing for Meaning

DISCUSSION QUESTIONS Have students read the Prologue and the Introduction to "Return to the Past." Invite student comments and questions on the Endangered Species Act. Allow about 10 minutes for students to complete the Task. Continue with a discussion of the following Analysis (also found in the Student Manual):

> **Prologue and "Return to the Past":** Teacher Guide pages 111-113.

- What resemblance do you see between the human population curve and the extinction graph?

- How are the extinctions caused by humans different from the ones discussed in Learning Experience 5? *(Students should respond that the others were mass extinctions possibly caused either by asteroids or by sudden and violent climactic changes, but recent extinctions seem to be linked to human behavior. Both in the past and in the present, there are naturally occurring gradual extinctions.)*

- What do you predict will happen to both human and other species populations in the future? Why do you think so?

Figure 6.1
Human behaviors that lead to extinctions. In many cases, two or more activities contribute to species extinction.

- What human behaviors are causing the acceleration of endangerment and extinction? (See Figure 6.1.)

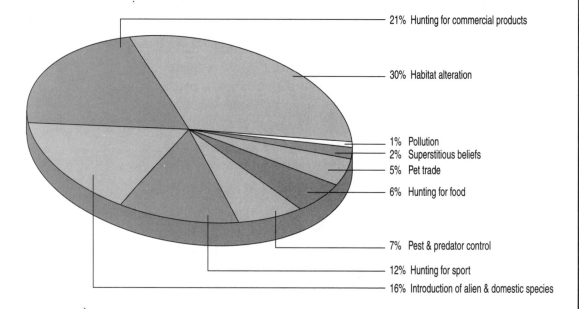

- 21% Hunting for commercial products
- 30% Habitat alteration
- 1% Pollution
- 2% Superstitious beliefs
- 5% Pet trade
- 6% Hunting for food
- 7% Pest & predator control
- 12% Hunting for sport
- 16% Introduction of alien & domestic species

"What Good Is It, Anyway?": Teacher Guide pages 114–115.

Continue the session by having students read "What Good Is It, Anyway?" Allow 10–15 minutes at the end of the session for students to create a "rainforest" concept map that shows the relationships within this ecosystem and the possible consequences from the destruction of the rainforest.

▶ **HOMEWORK**

"Save the *Fly*? Are you Kidding?": Teacher Guide pages 115–116.

Have students read the case study "Save the *Fly*? Are You Kidding?" and write an essay which details both the reasoning behind their opinion and the values that influenced their decision.

Applying
Session Two

DISCUSSION QUESTIONS Begin this session by having a few students share their concept maps. Invite students to comment on other connections and concepts that may be included. You may wish to elicit further discussion by asking questions such as the following:

- What unique characteristics make the rainforests important?
- Why do you think the rainforests are being destroyed at such a massive rate?

FOR FURTHER STUDY

- What are the economic, moral, political, and biological factors that should be considered when discussing the rainforest destruction?
- Is what is happening in Brazil important to the rest of the world? If so, why?

Continue the session by polling students as to their decisions about the Delhi Sands fly. Divide students into groups—those who would protect the fly, those who would not, and (if needed) those who are uncertain. Allow 15 minutes for each group to discuss their reasons and to choose two spokespersons.

DISCUSSION QUESTIONS Gather the class together for a discussion. Have the groups present their arguments for and against continued protection for the fly. They should also include the values that helped them come to such a decision. At the end of both presentations encourage students to comment on the opinions presented. You may wish to conclude the discussion with the following questions (modified from the Student Manual):

- Does protecting a species also mean protecting its habitat? Why or why not?
- What are the implications of habitat protection?
- Can one protect some species and not others? What would be the criteria?
- Would the criteria change depending on the species?
- What would be the basis for such a change?
- Does that mean that your values would change depending on the species? If so, how could you explain such a shift?

ASSESSMENT

▶ Have students made the connection between the species richness maps (in Learning Experience 6) and the tropical rainforest?

ASSESSMENT

▶ Have students shown evidence of careful and serious thinking about the case study?

▶ Are students' reasons logical and based on some scientific knowledge?

▶ Do students see the importance of protecting habitats in order to protect species?

For Further Study   Going...Going...Gone!

Going...Going... Gone!

PROLOGUE

Belayed 2,000 feet above Kauai's Kalalau Valley, botanists Ken Wood...and Steve Perlman risk their lives to rescue Hawaii's imperiled plants from extinction. Out of reach of goats and other alien invaders, such cliffs are among the last strongholds of the state's native flora. With the unhappy distinction of being the country's endangered species capital, the Hawaiian Islands have already lost hundreds of original life-forms while hundreds more teeter on the verge of oblivion.

(Excerpted from *On the Brink: Hawaii's Vanishing Species* by Elizabeth Royte, National Geographic, September 1995, p. 2.)

Why would anyone risk his or her life in an attempt to save a plant or animal species? Why are these organisms becoming endangered and going extinct? Does it really matter if they do? As you learned in Learning Experience 5, extinction is part of the natural cycle of life and plays an important role in evolution and continuing diversity of life on Earth. In this learning experience, you will explore some of the factors that result in the endangerment and loss of species and some of the factors that surround the preservation and protection of species.

Return to the Past

INTRODUCTION

The Endangered Species Act, passed by Congress in 1973 and reauthorized in 1988, regulates a wide range of activities which involve plants and animals designated as endangered or threatened. The Act defines an *endangered species* as a plant or animal that is rare and in immediate danger of extinction. A "threatened species" is any plant or animal which may become endangered in the

foreseeable future. The Act prohibits the following activities involving endangered species:

- importing them into, or exporting them from, the United States
- taking (including harassing, harming, pursuing, hunting, shooting, wounding, trapping, killing, capturing, or collecting) the species within the United States and its territorial seas
- taking them on the international high seas
- possessing, selling, delivering, carrying, transporting, or shipping any such species if unlawfully taken within the United States or on the high seas
- delivering, receiving, carrying, transporting, shipping them in interstate or foreign commerce in the course of a commercial activity
- selling or offering the species for sale in interstate or foreign commerce

Violators of this Act are subject to fines of up to $100,000 and one year of imprisonment. Organizations found in violation may be fined up to $200,000.

▶ TASK

1. Study Figures 6.3 and 6.4 on the facing page and note any relationships.
2. Write responses to the Analysis and be prepared to discuss your views.

▶ ANALYSIS

1. What resemblance do you see between the human population curve and the extinction graph?
2. How are the extinctions caused by humans different from the ones discussed in Learning Experience 5?
3. What do you predict will happen to both human and other species populations in the future? Why do you think so?
4. What human behaviors are causing the acceleration of species endangerment and extinction?

For Further Study

Notes

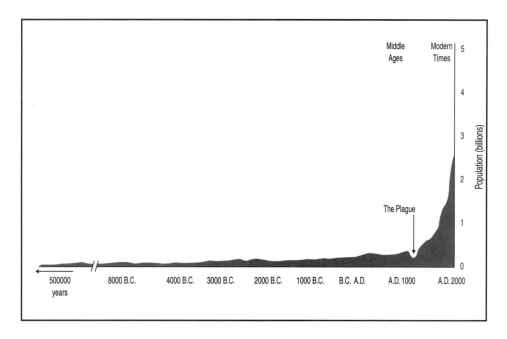

Figure 6.2
World population curve.

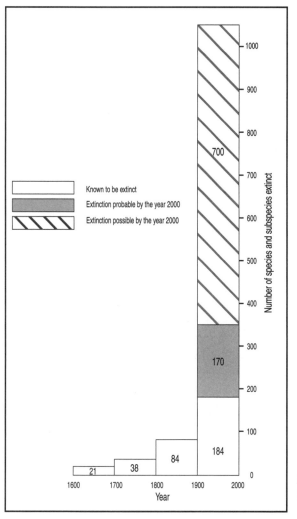

Figure 6.3
Number of vertebrate species lost.

For Further Study   Going...Going...Gone!

What Good Is It, Anyway?

Over the last 200 million years about 90 species per century, or less than one per cent per year, have become extinct. Judging from the fossil record, the natural average life span for an entire species is about 4 million years; if there are perhaps 10 million species on Earth today the extinction rate as a result of natural extinction would be about four species per year. According to some estimates, however, the annual rate of species loss may rise to 50,000 species per year by the year 2000.

Does it matter if a certain species becomes extinct? Are all species "equal" when thinking about this or are some more "valuable" than others? Who decides which to save and which to let go? What are the criteria for such a decision?

The rosy periwinkle (see Figure 6.4) has become a symbol of the untapped resources which lie within the dense growth of the rainforests and whose habitat is threatened by humans leveling rainforests. Found in the rainforests of Madagascar, this small plant with dark green leaves and a pink flower is a member of a large group of plants called *Vinca*. It has long been used in folk medicine for a range of ailments including dysentery, menstrual disorders, diabetes, and toothaches. Researchers have found substances in the roots, stems, and leaves of this plant which are extremely effective against two forms of cancer that were once fatal to children: Hodgkin's disease, a cancer primarily of young adults, and acute lymphocytic leukemia. The discovery of these substances (vinblastine and vincristine) in rosy periwinkle was a fortunate accident—a side result of the search for a cure for diabetes—but it has saved thousands of lives. Such discoveries enforce the concern that by deliberately or inadvertently destroying species, we may be destroying sources of products that could enhance or even save our own lives.

Figure 6.4
Rosy periwinkle.

The loss of the rainforest is of extreme concern to many people. The Amazonian rainforest in Brazil, for example, is a massive tropical ecosystem that covers five million square kilometers, making it the largest continuous expanse of tropical rainforest in the world. Although rainforests cover only seven percent of the planet's land surface, they contain 50 percent of the plants and animals found on the globe. The plants of the rainforest, whose potential medicinal value remain unknown, also serve as a source of oxygen for the planet's animal population.

FOR FURTHER STUDY

Many scientists seem resigned to the fact that most of the rainforests will eventually be cut down. At the present rate of destruction of 1,800 hectares (about 4,500 acres) per hour, this seems inevitable; E. O. Wilson calculates that 27,000 species are slated for extinction every year.

▶ **ANALYSIS**

1. Using "rainforest" as the central concept, create a concept map which shows the relationships within this ecosystem and the possible consequences from the destruction of the rainforest.

SAVE THE *FLY*? ARE YOU KIDDING?

"If you see one flying, you won't forget it. It's spectacular."

"I'm talking about jobs. How can you equate a fly with people's livelihoods?"

"When it's gone, it's gone forever. We want to save the habitat and the fly."

"The dinosaurs and the passenger pigeon are gone and I don't miss them."

What organism is causing such heated debate in San Bernadino County in southern California? It is the Delhi Sands flower-loving fly, which, in 1993, became the first fly to be put on the endangered species list for protection. There are so few left that they can almost be individually counted.

Delhi Sands is the largest remaining inland sand dune system in the Los Angeles basin. Its habitat supports not only the fly but pocket mice, butterflies, an unusual cockroach-like cricket, and rare flowers. Of the original 40 square miles, 300 acres are available for the fly's needs and, in effect, only 45 acres actually support it. It appears to be heading quickly for the same fate—extinction—as its sister subspecies, the El Segundo fly, buried under the tarmac of the Los Angeles International Airport.

Figure 6.5
Delhi Sands fly.

Many factors have led to the loss of this fly's habitat and to its reduction in numbers, including construction, off-road vehicles, spraying for weeds, bulldozing for fill, and the dumping of manure from local dairies. Because the fly is protected, a hospital under construction has been moved a few hundred feet, construction of a subdivision has been stalled, and plans for a massive industrial development has been halted with a potential loss of thousands of jobs. A lawsuit filed to end

the federal government's protection of the fly was dismissed but has been appealed.

The obscure fly spends most of its life underground during its larval stage as a maggot. The entomologists who study the fly are not sure what the maggots eat. The adults emerge from the earth in the summer and fly for a few days, hovering like hummingbirds above flowers and extracting their nectar with a long straw-like mouthpart. (Such a structure is an example of adaptive evolution.) The males fly over the dunes in the heat of the day looking for females with which to mate. No one is sure how the flies mate, but the females then lay fertilized eggs deep in the sand; the eggs hatch in a week. Entomologists consider the Delhi Sands fly an interesting creature in an important place, each worthy of protecting.

The debate on saving species seems less controversial when it involves species appealing to humans, such as cuddly-looking baby seals, Bambi-like deer, or songbirds. But this fly, like the snail darter and the northern spotted owl before it, is generating political, moral, and biological debate. On one side are those who want to weaken the Endangered Species Act, and who see the protection of this supposedly insignificant creature as an example of the absurdity of saving all endangered species. On the other side are entomologists (those who study insects) and others who support the Act to its fullest, that is, protecting creatures great and small. Is there a middle ground? Can one support the protection of some organisms, but not others? What would be the criteria? Would the criteria change depending on the species? What would be the basis for such a change in view?

What is your reasoned opinion? Should the Delhi Sands fly be protected? Why or why not? What values do you have that support your decision? Does your view hold true for all endangered species? Why or why not?

Back to Nature

OVERVIEW By some accounts, humans have disturbed or altered more than half of the world's terrestrial ecosystems through pollution, environmental degradation, resource depletion, and population pressures. What does this mean for humanity? In this learning experience, students will explore in greater depth the intersection of the actions of humans and other species. Students have been studying the interconnections of organisms and the relationship of organisms to the abiotic factors in the environment. The basic principles that govern the interactions of other organisms affect human interactions as well; yet humans often act as if they can alter the environment with no consequences. Can the damage humans have done be repaired?

In this learning experience, students will explore how advances in ecological understandings may allow ecosystems damaged by human actions to be repaired. Specifically, students will look at the plans to restore the Florida Everglades and at the issues that surround the restoration of this area, and decide what they would do.

LEARNING OBJECTIVES

- *Students apply principles of ecology to analyze an ecosystem restoration project.*
- *Students research human needs versus the goal of preserving species and ecosystem diversity in a unique ecosystem—the Everglades.*
- *Students evaluate the pros and cons of a restoration ecology project in the Everglades and determine their own views.*

▶ SUGGESTED TIME

- 5–6 class sessions (45–50 minute periods)

▶ ADVANCE PREPARATION

1. Read through the learning experience and determine whether your students will do the research project on the Everglades or the research project described in the teaching strategy on pages 120–121 of this learning experience. The latter project encourages students to research a restoration project that is occurring in their own city or state. If you choose to have students do this research project, you may need to help students identify several restoration projects and contacts.

2. Alert the personnel in the media and/or computer centers to the needs of your students regarding access to the World Wide Web and to other resources on the Everglades Restoration Project. You may want to give the staff a copy of the Resource list for this learning experience.

Learning Experience 7 Back to Nature 117

3. Locate a large map of Florida and the Everglades and post it in your classroom for the duration of this project.
4. Invite colleagues, administrators, or other students into the class to be members of the mock congressional committee in Session Five or Six.

▶ ASSUMPTIONS OF PRIOR KNOWLEDGE AND SKILLS

- Students are familiar with using a library and/or the World Wide Web for conducting research.

▶ TEACHING SEQUENCE PREVIEW

SETTING THE CONTEXT

- Students explore the new field of restoration ecology and discuss the ecological principles that should be considered when attempting to restore ecosystems.

EXPERIMENTING AND INVESTIGATING

- Students are introduced to the research project "Anything We Want" which has them identify the issues surrounding the Everglades Restoration Project and research one issue in depth.

PROCESSING FOR MEANING

- Students deliver speeches detailing the issues they researched to a mock congressional committee.
- Students discuss the pros and cons of the Everglades Restoration Project based on the presentations and write a paper detailing what they would do if they were members of the committee.

TEACHING SEQUENCE

Setting the Context

Session One

Begin class by having students read the Prologue, which introduces restoration ecology. Have a class discussion which focuses on general ecological principles that should be considered when restoring ecosystems or habitats.

Prologue: Student Manual page 71.

DISCUSSION QUESTIONS In this Setting the Context, students relate the ways in which human actions have altered or diminished ecosystems to the principles of balance in ecosystems that they have been exploring in the module. To begin the discussion you may wish to ask:

- What is the purpose of a restoration project?
- What are some examples of ecological restoration? *(Student responses may include dam removals in certain rivers, the reintroduction of endangered species such as the condor, prairie restoration in the Midwest, pollution cleanup in toxic dumps, or other local projects.)*
- Why is a restoration project thought to be necessary? *(Student responses will vary depending on the type of restoration project they are discussing. This is true for responses to the remaining questions.)*
- What are the pros and cons of restoring land to its original conditions?
- Should humans interfere or should they allow natural forces to restore the environment?
- In what ways might a restoration project affect the species that currently live in an area?
- In what ways might a restoration project affect human populations?
- Who should pay for restoration projects?

TEACHING STRATEGY

You may also wish to mention to your students that other, perhaps less complex initiatives, such as landscape ecology and conservation ecology, also attempt to preserve ecosystems.

Experimenting and Investigating

In the activity "Anything We Want," students read about the Florida Everglades and the attempt to restore the wetland ecosystem. Students will identify issues that surround this project and research one issue in greater depth.

"Anything We Want": Student Manual pages 72–78.

Have students read the Introduction and Task to "Anything We Want." Review the main points and your expectations for this activity. Inform students that during the next few sessions (whatever time frame you feel is appropriate), they are to research their issue and develop their presentations.

Divide the class into pairs and have the pairs begin step 1 of the Task. Each student should be recording the list in his/her notebook.

▶ HOMEWORK

Have each student complete step 1 of the Task.

Session Two

Allow each pair 5–10 minutes to compare their completed lists of the issues and the involved groups.

DISCUSSION QUESTIONS Gather the class together and have volunteers state the issues. Record a list on the board or chart paper. After the list is complete, you may wish to discuss the main points of the Everglades Restoration Project and ask the following:

- What is the purpose of the Everglades Restoration Project?
- Who are the individuals, organizations, or groups involved in the Project?
- What are the major categories of issues surrounding the project? *(Student responses should include the ecological/scientific unknowns, the social tension between population growth into the flood plains and the desire to control floods [including need for housing developments, and the development of roads and golf courses], the continually changing political environment, and the economic stresses placed on the sugar farmers.)*

TEACHING STRATEGY

During the remaining sessions of this learning experience, students will engage in a research project on the Florida Everglades. If you feel that this research project may be difficult for your students because of lack of resources or access to the World Wide Web, you may wish to have them do the following activity, which asks students to research a restoration project that is occurring in their area.

Restoration ecology, landscape ecology, and conservation-biology are in many areas of the United States and even of the world. People are rebuilding rivers, streams, ponds, and beaches; reconstructing forests and prairies and deserts; coaxing populations of near extinct species back to size. Have students research a restoration project that is occurring in their area. You may wish to help them identify projects that are occurring and ways in which to go about finding out information on the projects. This could include library research as well as visiting sites and writing to those involved in the project.

Have students write a report or develop a brochure which describes the project; why there was a need for the project; how it was funded; the habitats/species that were affected; the groups, companies, or individuals that are responsible/or affected by the project; any controversy or issues that resulted from the projects; and the successes and failures of the project.

After this discussion, have pairs identify an issue that they would like to research. Try to have each pair research a different issue. Explain that the next few sessions are set aside for students to do research in the media and/or computer center. Since this project is a current one, it is recommended that the groups use the World Wide Web as they search for newspaper, magazine, and scientific articles on the topic. Although computer access is not required for this project to be successful, it would enrich the information flow and enhance the quality of the speeches.

Student analysis of the article and further research should reveal the following ecological and environmental issues:

- There is no evidence that a created wetland can match the richness of one that is destroyed. There are 55 endangered species at risk.
- Scientists disagree on the process of restoration and cleaning the water in Florida.
- The political climate in the country and in Florida changes regularly, which puts sustained funding at risk.
- An increase in pressure on agriculture and industry could mean a loss of profits and jobs in the area.
- The increasing population of Florida puts pressure on land and water use. Cities face water shortages.
- The cost of inaction is the loss of biological diversity and a unique ecosystem.
- Hydrologic restoration does not necessarily equal biological restoration.
- Restoration and clean-up would require higher taxes to pay for the change in the infrastructure and higher water bills.

- As the individuals, groups, and institutions that support restoration represent a broad coalition of interests, staying united may be difficult.
- The restoration will cost a lot of money, and not everyone agrees the restoration will work. Some feel that restoration could even promote more destruction because if people believe nature can be rebuilt there is no harm in losing more of the ecosystem. Others feel restoration is the only possible way of responding to the damage.

THINGS TO WATCH FOR

Although the World Wide Web is a rich source of information, it can also be misleading, or even a rumor mill. Caution students—and be wary yourself—about examining the source of any information obtained in this way.

▶ HOMEWORK

Encourage students to begin research on the issue they have chosen and to be prepared to spend the next few days in class continuing and synthesizing their research and developing their speeches.

Sessions Three through Four or Five

Have students continue researching their topics and developing their speeches. They will need your assistance as they work. Circulate and ask questions such as the following:

- What are some of the current ideas and concerns that surround your issue?
- Do you agree or disagree with the concerns raised?
- What questions do you have about the restoration project?
- How can you find the answers?

▶ HOMEWORK

Students should continue either researching the topic or developing their presentation.

Processing for Meaning

Session Five or Six

Begin the session by reviewing the goal of the congressional hearing. Explain that you will be a member of the committee, and introduce the other members of the committee (see Advance Preparation). Remind students that they are to listen carefully to other presenters and take notes on the speeches. This is important because following the presentations there will be a discussion which focuses on the questions raised and presented by the research of each group. In addition, the committee members may ask questions of the researchers to clarify points raised, and thus, help students make their decisions about the fate of the Everglades.

▶ HOMEWORK

Have students complete step 5 of the Task in which students write a comprehensive paper summarizing the rationale of the project and all sides of the issues heard at the presentations. They then are to describe their own opinion about what should be done in the Everglades complete with reasons based on the presentations.

Students' position papers should include and be evaluated on the following:

- a rationale for the development of the restoration project
- the ecological principles from this module that apply to the restoration project such as:
    - how human interference in an ecosystem affects its functioning
    - how the diversity of species in this unique ecosystem results from evolutionary forces
    - that restoration can not necessarily recreate what existed
    - the role of pockets of native diversity in repopulating restored areas
- a clear description of many of the issues listed above
- an educated opinion about what should be done

Students' papers should be written clearly, detail all the issues, and offer an educated opinion that follows from logical reasoning.

Appendices

Appendices

Materials List

▶ MATERIALS NEEDED BY LEARNING EXPERIENCE

(Amounts listed are enough for 32 students/8 groups of 4 students or 16 pairs when applicable.)

LEARNING EXPERIENCE 1
- 8 large bags or boxes to carry materials
- 8 meter sticks or tape measures
- 32 large nails
- 8 hammers, mallets, or fist-sized stones
- 40 m of string
- 8 thermometers
- 8 trowels, spades, 5-cm diameter piece of pipe, or small bulb planters
- 8 sheets of graph paper
- 8 sheets of newspaper
- 16 hand lenses
- 8 petri dishes
- 1 or more soil test kits (optional)
- 16 or more 2-L bottles of the same brand with caps
- 8 wax marking pencils
- 8 single-edge razor blades or scalpels
- 8 scissors
- 8 awls or nails
- clear waterproof tape
- 8 darning needles or large safety pins
- silicone sealant
- soil, water, plants, compost, fruit flies, spiders, snails and/or other small organisms
- wick (cotton rope or other absorbent material)
- 2 or 3 large shoebox tops

LONG-TERM PROJECT
- soil test kits including:
  - nitrogen, phosphorous, and potassium test kits
  - acidity and alkalinity test kits
  - microbe study kit
  - microorganisms test kit
- pH meter

Appendix A   Materials List   **125**

- water test kits including:
  - algae pollution kit
  - bacteria pollution kit
  - coliform pollution kit
  - quantitative analysis kit
  - dissolved oxygen water test kit
  - acid rain survey kit
  - water filter kit
- water hardness kit
- seawater analysis kit
- limnology water test kit
- lead testing kit
- cameras and film
- sketch pads
- tape recorder
- hand lenses
- jars and bottles
- field guides
- rain gauge
- thermometers

LEARNING EXPERIENCE 2

- 8 zipper plastic bags each containing a sample of soil from one of three or four different locations (see Advance Preparation)
- 16 hand lenses
- 16 tweezers
- 16 pointers (or coffee stirrers)
- 8 sheets of newspaper
- 8 petri dishes
- dissecting microscope
- field guides (plants, insects, other invertebrates)
- 8 rotting logs (if available)
- 17 biota cards
- 1 cardboard "sun" (see Advance Preparation)
- 1 ball of yarn or string
- 1 scissors
- cellophane tape
- 1 meter stick or other pointer
- invertebrate field guides (optional)
- 5-cm diameter piece of pipe, small bulb planter, trowel or spade

LEARNING EXPERIENCE 3

- 32 sheets of unlined white paper (11 x 14 inches, legal size)
- felt-tip markers or colored pencils
- 32 pairs of safety goggles
- 32 culture tubes or test tubes with tight-fitting covers or stoppers
- 8 test tube racks
- 8 wax marking pencils
- 1–2 L distilled water (or "aged" tap water)
- 8 eyedroppers

- 8–16 mL 0.1% bromothymol blue solution (aqueous)
- 32 small water (pond) snails
- 16 sprigs of Elodea
- access to a fluorescent lamp
- 4 transparency sheets
- salt (optional)
- hot water (optional)
- 1 plastic bag (optional)
- 1 spoon (optional)
- ice cubes (optional)
- 1 2-L plastic bottle with cap (optional)
- 1 1-L plastic bottle with cap (optional)
- 2 laboratory stands (optional)
- 2 bowls (optional)

LEARNING EXPERIENCE 4

- 24 sheets of graph paper
- 800 hare cards (5 cm x 5 cm) (See Advance Preparation)
- 200 lynx cards (10 cm x 10 cm) (See Advance Preparation)
- 8 rulers or tape measures
- masking tape

FOR FURTHER STUDY OF LEARNING EXPERIENCE 4

- 64 sheets of graph paper
- cellophane tape
- 1 knife (optional)
- 1 apple (optional)

LEARNING EXPERIENCE 5

- 16 "mystery" objects (such as: potato ricer, strawberry huller, olive pitter, dandelion digger, hose nozzle, nutcracker, hardware item, etc.)
- 16 paper bags (lunch size)
- 32 sheets of graph paper
- 32 colored pencils
- label or index card

LEARNING EXPERIENCE 6

- 10 sets of "Species Richness" maps
- assorted colored pencils
- reference material including atlases, encyclopedias, and geography books
- 2 or 3 transparency sheets (optional)

FOR FURTHER STUDY OF LEARNING EXPERIENCE 6
none

LEARNING EXPERIENCE 7
none

Resource List

BOOKS BY LEARNING EXPERIENCE

Learning Experience 1

Bauman, Edward. *The Essential Aquarium: A Guide to Keeping 100 Exciting Freshwater Species.* New York: Crescent Books, 1991.

Blasiola, George C., II. *The New Saltwater Aquarium Handbook.* New York: Barron's, 1991.

Bonta, Marcia Myers. *Women in the Field: America's Pioneering Women Naturalists.* College Station, TX: Texas A & M University Press, 1991.

Faber, Doris and Harold Faber. *Nature and the Environment.* New York: Charles Scribner's Sons, 1991.

Feltwell, John. *The Naturalist's Garden.* Topsfield, MA: Salem House, 1987.

Golden Guide. Titles include: *Pond Life; Spiders; Butterflies and Moths; Birds; Flowers; Insects; Trees; Reptiles and Amphibians; Mammals; Fish; Bird Life; Weeds; Seashores;* and *Endangered Species.*
A series of easy-to-use field guides.

Grossman, Shelly. *Understanding Ecology.* New York: Grosset & Dunlap, 1970.
The author gives a grand portrayal of the many aspects of the North American continent. Beautiful photographs accompany fascinating text which describes in great detail the plants and animals, and the physical environments they inhabit on this continent.

Herda, D.J. *Making a Native Plant Terrarium.* New York: Messner, 1977.

Lawrence, Gale. *The Indoor Naturalist: Observing the World of Nature Inside Your Home.* New York: Prentice Hall Press, 1986.
Origins, life cycles, and interactions with the various other organisms that can inhabit a human household are described in intricate detail as the author explores the fascinating ecology of animals, plants, and household products inside the home.

McNee, Thomas. *The Return of the Wolf to Yellowstone.* New York: Holt, 1997.

National Council for Agricultural Education. *Using Fast Plants and Bottle Biology in the Classroom.* Reston, VA: National Association of Biology Teachers, 1994.

> A helpful activity book suggesting ideas for testing numerous factors for their effects on plant growth.

Learning Experience 2

Carson, Rachel. *Silent Spring.* Boston, MA: Houghton Mifflin Company, 1962.

> This is the book that changed America's way of thinking about DDT spraying in particular, and raised our awareness of the environment in general.

Hingley, Marjorie. *Fieldwork Projects in Biology.* Poole, Dorset, England: Blandford Press, 1979.

> This book contains many open-ended fieldwork projects covering a wide range of habitats, which can be used for long term or short term projects, or as a source book.

Jezer, Marty. *Rachel Carson.* American Women of Achievement. New York: Chelsea House Publishers, 1988.

> A biographical account of the marine biologist and author whose writings stressed the interrelationships of all living things and the dependence of human welfare on natural processes.

White, William, Jr., Ph.D. *Forest and Garden. Living Nature Series.* New York: Sterling Publishing Co., Inc., 1976.

> This book describes the natural processes that create and maintain the forest and the artificial techniques by which man creates and maintains the garden, and examines key life forms to be found in each.

Whitfield, Philip, Peter D. Moore, and Barry Cox. David Attenborough: *The Atlas of the Living World.* Boston, MA: Houghton Mifflin Company, 1989.

> A magnificent account of the patterns of life on earth, this book focuses on the various environments, biological processes, and the natural history of the planet. Through photos, charts and graphs, and informative text, the unique identity and the dynamic nature of each of the plants, animals, and habitats is majestically revealed.

For Further Study of Learning Experience 4

Cohen, Joel E. *How Many People Can the Earth Support?* New York: W. W. Norton & Company, 1995.

> A deep analysis of this not so simple question, this book explores the limits of both natural and human-induced constraints, a review of the history of world population growth, and objectively identifies strategies for dealing with some of the fundamental problems.

Commoner, Barry. *Making Peace with the Planet.* Raytheon Press, 1990.

Ehrlich, Paul R. *The Population Bomb*. New York: Ballantine Books, 1968.
>Mr. Ehrlich's original book describing his view of runaway human population growth and his dire predictions foe the outcome of the world if such growth continues.

Ehrlich, Paul R. and Anne H. Ehrlich. *The Population Explosion*. New York: Simon & Schuster, 1990.
>From global warming to rain forest destruction, famine, and air and water pollution, the authors expound on why overpopulation is our number one environmental problem.

Ehrlich, Paul R., Anne H. Ehrlich and Gretchen C. Daily. *The Stork and the Plow–the Equity Answer to the Human Dilemma*, G. P. Putnam's Sons, 1995.

Learning Experience 5

Darwin, Charles. *The Origin of Species and The Descent of Man*. New York: Random House.
>Darwin's original writings of his observations of plants and animals around the world during his five-year voyage on the H.M.S. Beagle.

Darwin, Charles. *The Voyage of the Beagle*. New York: Dutton Publishing, 1965.
>Mr. Darwin's accounts of his observations during his five-year sea voyage, as recounted in his journals.

Dawkins, Richard. *The Blind Watchmaker*. New York: W.W. Norton & Company, 1986.

Dixon, Desmond. *After Man: A Zoology of the Future*. New York: St. Martins Press, 1981.

Gould, Stephen Jay. *Ever Since Darwin*. New York: W. W. Norton & Company, 1977.
>This interesting and informative book reprints some of Gould's columns which originally appeared in Natural History Magazine.

Gould, Stephen Jay. *Wonderful Life*. New York: W. W. Norton & Company, 1989.
>This tale of the fossil remains of extinct sea species trapped in a limestone quarry high in the Canadian Rockies over 350 million years ago (the Burgess Shales) is a fascinating story of evolution preserved.

Ward, Peter. *The End of Evolution: On Mass Extinctions and the Preservation of Biodiversity*. New York: Bantam Books, 1994.
>The author recounts stories and studies made around the world of past and recent signs of extinctions of various species.

Weiner, Jonathan. *The Beak of the Finch: A Story of Evolution in Our Time*. New York: Alfred A. Knopf, 1994.

> The story of two evolutionary biologists on the Galapagos Islands who are watching and recording evolution as it is occurring now, among the finches that inspired Charles Darwin. They are studying the evolutionary process in real time, in the wild.

Learning Experience 6

Baskin, Yvonne. *The Work of Nature: How the Diversity of Life Sustains Us*. Washington, D.C.: Island Press, 1997.

> The author systematically details the factors of soil, water, and plants all benefit from biologically diverse ecosystems and how all of them, in turn, benefit humans. She assesses the loss of diversity by inventorying several ecosystems and noting their interdependencies.

Mann, Charles C. and Mark L. Plummer. *Noah's Choice: The Future of Endangered Species*. New York: Alfred A. Knopf, 1995.

> A very well researched book that looks at endangered species, the authors provide thoughts about the responsibility of choosing among species and striking balance between the needs of human beings and the rest of the world.

Wilson, Edward O. *Biodiversity*. Washington, D.C.: National Academy Press, 1988.

> A compilation of essays by scientists, economists, agricultural experts, philosophers, and other experts who discuss the loss of habitats and biodiversity from their perspectives.

Wilson, Edward O. *Naturalist*. Washington, D.C.: Island Press, A Shearwater Book, 1994.

> Recounting his vast and varied experience in studying the natural world, Wilson alerts the reader of the plight of organisms around the planet and describes their interrelationships with humans, in this, his autobiography.

Wilson, Edward O. *The Diversity of Life*. Cambridge, MA: The Belknap Press of Harvard University Press, 1992.

> In this important book, Wilson tells the story of how life on Earth evolved and how the species of the world became diverse, and explains why the threat to that biodiversity today is serious and monumental in scope and implication.

For Further Study of Learning Experience 6

Kreig, Margaret B. *Green Medicine: The Search for Plants that Heal*. Chicago, IL: Rand McNally & Company, 1964.

> Intriguing true stories of discoveries from botanical investigations which have resulted in some of our most important medicines. This scientifically accurate, as well as entertaining, book is largely based on field journals and personal recollections of the scientists themselves who made the discoveries.

Plotkin, Mark J. *Tales of a Shaman's Apprentice*. New York: Viking Penguin, 1993.

> An ethnobotanist searches for new medicines in the Amazon rain forest.

Stalcup, Brenda, Ed. *Endangered Species: Opposing Viewpoints.* San Diego, CA: Greenhaven Press, Inc., 1996.

> Another set of essays in the interesting opposing viewpoints series, issues discussed include endangered species and biological diversity conservation in the United States and worldwide.

Learning Experience 7

Berger, John J. *Environmental Restoration: Science and Strategies for Restoring the Earth*, Island Press, 1990.

Davis, Stephen M. and John Ogden. *Everglades: The Ecosystem and Its Restoration.* Delray Beach, FL: St. Lucie Press, 1994.

Dobson, Andrew P. *Conservation and Biodiversity.* New York: Scientific American Library, 1996.

> Using discussions of species diversity, extinction rates, and the process of determining species as endangered, Mr. Dobson is also attentive to the economics of species conservation and management, and describes and analyzes strategies for making preservation viable.

Douglas, Marjory Stoneman. *The Everglades: River of Grass.* New York: Rinehart & Company, Inc., 1947.

> Written in an engaging narrative format, Ms. Douglas describes in detail the Florida Everglades, using her perception and eloquent writing skills to help illustrate and preserve the scientific importance of the Everglades.

Douglas, Marjory Stoneman. *Voice of the River. An Autobiography with John Rothchild.* Englewood, FL: Pineapple Press, Inc., 1987.

> A fascinating autobiography of the extraordinary woman who "saved" the Everglades, this book gives a delightful insight to her century-long life and recollections.

Heuvelmans, Martin. *The River Killers.* Harrisburg, PA: The Stackpole Company, 1974.

> A painstakingly researched book, the author gives detailed accounts documenting misrepresentations and mismanagement by the Civil Works Project Branch of the U. S. Army Corps of Engineers, focusing mainly on their work in Florida.

McPhee, John, *The Control of Nature.* New York: Farrar Straus Giroux, 1989.

Restoration of Aquatic Ecosystems: Science Technology, and Public Policy. National Research Council. National Academy Press, 1992.

Zim, Herbert S., with the cooperation of Florida National Parks and Monuments Association, Inc. *A Guide to Everglades National Park and the Nearby Florida Keys.* New York: Golden Press, 1992.

> An extensive pocket guide book which describes the flora and fauna of the area, the history of the people and the land, as well as what to see and do around the park areas.

GENERAL BOOKS

Asimov, Isaac and Frederik Pohl. *Our Angry Earth*. New York: Tom Doherty Associates, Inc., 1991.

> The authors use the most recent scientific information available to provide a view of the current state of our planet and an overview of the reforms that need to be instituted for the health of the environment.

Berkman, Richard L. and W. Kip Viscusi. *Damming the West*. New York: Grossman Publishers, 1973.

> Ralph Nader's study group report on the Bureau of Reclamation (set up by Congress as an irrigation agency responsible for reclaiming/irrigating the arid lands of the American West).

Brown, Lester, et al. *State of the World: A Worldwatch Institute Report on Progress Toward a Sustainable Society*. New York: W.W. Norton, annually since 1984.

> Published each year, these volumes offer a unique perspective analyzing the ecology and economy of the progress toward a sustainable global society.

Cunningham, William P. and Barbara Woodworth Saigo. *Environmental Science: A Global Concern*. Dubuque, IA: Wm. C. Brown Communications, 1995.

> A text focusing on the global scope of environmental science and its importance for an acceptable quality of life.

Dillard, Annie. *Pilgrim at Tinker Creek*. New York: Harper's Magazine Press, 1974.

> A gifted writer shares her observations and insights about the natural world in this classic, prize-winning book.

Earthworks Group, The. *50 Simple Things Kids Can Do to Save the Earth*. Kansas City, MO: Andrews and McMeel, 1990.

> An interesting how-to book filled with experiments, facts and things to do.

Goldfarb, Theodore D. *Taking Sides: Clashing Views on Controversial Environmental Issues*. Guilford, CT: The Dushkin Publishing Group, Inc., 1987.

> Looking at opposing viewpoints and arguments on several environmental issues.

Gonick, Larry and Alice Outwater. *The Cartoon Guide to the Environment*. New York: Harper Perennial, 1996.

> Using lively, witty cartoon illustrations along with descriptive, interesting text, the authors cover the main topics of environmental science and put them into the context of ecology, including discussions of the behavior of complex systems, population dynamics, and thermodynamics.

Gore, Al. *Earth in the Balance*. New York: Houghton Mifflin Company, 1992.

Heloise. *Heloise Hints for a Healthy Planet.* New York: Perigee Books, 1990.
> A collection of hints and tips for reusing, recycling, or otherwise saving resources when doing jobs in and around the home, yard, and office.

Mann, John. *Murder, Magic and Medicine.* New York: Oxford University Press, 1994.
> Descriptions and histories of numerous drugs, links between the folk use of plant and animal extracts and their modern use as drugs.

McCuen, Gary E. and David L. Bender, ed. *The Ecology Controversy: Opposing Viewpoints.* Minnesota: Greenhaven Press, 1973.
> Sets of essays providing arguments for and against such ecological issues including population growth, Western culture's effects on the environment, and human values.

McPhee, John. *Encounters With the Archdruid: Narratives about a Conservationist and Three of his Natural Enemies.* New York: Farrar Straus and Giroux, 1971.
> Mr. McPhee, in a wonderful narrative style, has recorded stories of some particular people involved in using the environment for different purposes, including a mineral engineer, a resort developer, a dam builder, and a militant conservationist.

Quammen, David. *The Song of the Dodo: Island Biogeography in an Age of Extinctions.* New York: Scribner, 1996.
> The author explores the question of why island species face such high rates of extinction and looks at the conditions leading to their disappearance.

Stevens, William K. *Miracle Under the Oaks: The Revival of Nature in America.* New York: Pocket Books, a division of Simon & Schuster Inc. 1995.
> The first book to look at the momentous new grassroots environmental restoration movement, the challenge to figure out how to revive and restore particular ecosystems is enthusiastically taken on and documented.

ARTICLES (FROM MAGAZINES AND NEWSPAPERS)

Learning Experience 1

Bicak, Charles J. The Application of Ecological Principles in Establishing an Environmental Ethic. *The American Biology Teacher.* April 1997: 200-206.

Doctors of the Wilderness. *Natural History*, May 1992, vol. 101: 42 (4).

Ethical Issues in the Release of Animals from Captivity. *Bioscience,* February 1997, vol. 47: 115 (7).

Radetsky, Peter. Back to Nature. *Discover Magazine.* July 1993: 34-42.

Stolzenburg, William. Building a Better Refuge. *Nature Conservancy,* January/February 1996.
> Habitat architects are learning more about combinations of factors in ecosystems, how they interact to provide suitable environments, and are designing new nature reserves as a result.

Long-Term Project

Friday, Gerald. Environmental Niches. *The Science Teacher,* May 1997, vol. 64: 38–41.

Lee, Judy and Joyce DeRulle. Field-Testing Ozone. *The Science Teacher*, December 1995, vol. 62: 16–19.

Singletary,, Ted J. and J. Richard Jordan. Exploring the GLOBE. *The Science Teacher*, March 1996, vol. 63: 36–39.

Learning Experience 2

Ford, Bob and Bruce M. Smith. Food Webs and Environmental Disturbance: What's the Connection? *The American Biology Teacher,* April 1994, vol. 56: 247–249.

Learning Experience 3

Chandler, David. Food Was an Obsession In Biosphere 2. *The Boston Globe*, October 4, 1993.
> Example of how food is grown and used in Biosphere 2, a completely separate, self-contained system, and some of the complications.

Schipper, Angela, Louis Schipper, and Art Hornsby. The Nitrogen Cycle. *The Science Teacher*, April 1996, vol. 63: 34–37.

Learning Experience 4

Bright, Chris. Bio-Invasions: The Spread of Exotic Species. *World Watch*, July/August 1995.
> Exploring the issue of different species moving into new areas, often to the detriment of indigenous species, and how that affects global diversity and economics.

Cohen, Joel. Ten Myths of Population. *Discover Magazine*, April 1996: 42–47.

Earth Grows More Crowded. GeoPlus Main News, *Weekly Reader,* November 18, 1994.
> Population growth statistics and ideas for growing new crops and crops in new places, such as in extreme cold climates and within cities.

For Further Study of Learning Experience 4

Brown, Lester R. Facing Food Scarcity. *World Watch*. November/December 1995: 10-20.

Hinrichsen, Don. Putting the Bite on Planet Earth. *International Wildlife*, September/October 1994.

Raloff, Janet. The Human Numbers Crunch. *Science News*. June 22, 1996, vol. 149: 396-397.

Learning Experience 5

Alters, Brian J. and William F. McComas. Punctuated Equilibrium: The Missing Link of Evolution Education. *The American Biology Teacher,* September 1994, vol. 56: 334–340.

Diamond, Jared. How to Tame a Wild Plant. *Discover Magazine,* September 1994.
Looking at modern crops and how they evolved from wild plants.

Gould, Stephen Jay. Darwin and Paley Meet the Invisible Hand. *Natural History,* November 1990: 8-16.

Gould, Stephen Jay. The Tallest Tale. *Natural History.* May 1996.
In this essay, Gould suggests that the traditional teaching of the evolution of the giraffe neck may be a "bit of a stretch."

Keown, Duane. Teaching Evolution: Improved Approaches for Unprepared Students. *The American Biology Teacher*, October 1988, vol. 50: 407–410.

Miller, Kenneth R. Life's Grand Design. *Technology Review*, February/March 1994.

National Association of Biology Teachers. NABT Unveils New Statement on Teaching Evolution, March 15, 1995. *The American Biology Teacher*, National Association of Biology Teachers, 1995.

Learning Experience 6

Goodman, Billy. Drugs and People Threaten Diversity in Andean Forests. *Science*, American Association for the Advancement of Science, June 1993.

Harte, John. Defining the 'B' Word. *Defenders,* Spring 1996.
The author describes various factors in the three general kinds of biodiversity: species diversity, genetic diversity, and habitat diversity.

Monastersky, R. Minor Climate Change Can Unravel a Forest. *Science News*, November 27, 1993.
A modest climate cooling several hundred years ago upset the balance of tree species inhabiting southern Canada, suggesting that even changes spread over several centuries can dramatically alter forests and reduce their productivity.

Nadis, Steve. Biodiversity for Sale. Continuum, *Omni,* March 1995.

For Further Study of Learning Experience 6

Booth, William A. A Fly in th Ointment *The Washington Post National Weekly Edition.* April 14, 1997: 9.

Mann, Charles C. and Mark L. Plummer. The Butterfly Problem. *The Atlantic Monthly.* January 1992: 47–70.

McAuliffe, Kathleen. Shaman Pharmaceuticals. *Omni*. July 1993: 27.

Pennisi, Elizabeth. Tallying the Tropics: Seeing the Forest through the Trees. *Science News*, June 4, 1994.
> Monitoring species diversity for more than a decade, scientists watch trees grow and produce fruit, and monitor seedlings in order to learn what controls the organization of the forest.

Peters, Robert L. and Reed F. Noss. America's Endangered Ecosystems. *Defenders of Wildlife*, Fall 1995.
> Issues and descriptions of the top 21 endangered ecosystems, due to human development, the types of development and the effects.

Winkler, Suzanne. Stopgap Measures. *The Atlantic Monthly*. January 1992: 74-81.

Zimmer, Carl. More Productive, Less Diverse. Ecology Watch section, *Discover Magazine*, September 1994.

Learning Experience 7

Adler, Tina. Bringing Back the Birds. *Science News,* August 17, 1996, vol. 150: 108-109.

Anderson, William Truett. There's No Going Back to Nature. *Mother Jones*, September/October 1996: 34-37, 76-79.

Barinaga, Maria. A Recipe for River Recovery? *Science,* September 20, 1996, vol. 273: 1648-1650.

Derr, Mark. Redeeming the Everglades. *Audubon,* September/October 1993: 48-131 (12).
> A description of what has happened to the Everglades since humans tried to dry it up, then tried to return it to its natural state.

Hanson, Beth. Blockbuster Battle. Wetlands Debate section, *Science World,* October 7, 1994.
> Will a planned Blockbuster theme park endanger Florida's remaining wetlands?

Holloway, Maguerite. Nurturing Nature. *Scientific American,* April 1994: 98-108.

Homer-Dixon, Thomas, et al. Environmental Change and Violent Conflict. *Scientific American*, February 1993.

Improving the Success of Wetland Creation and Restoration and Know-How, Time, and Self-Design. *Ecological Applications,* February 1996, vol. 6: 77 (7).

Levin, Ted. Immersed in the Everglades, *Sierra,* May/June 1996.

Louma, Jon R. Soiling the Planet. *Discover Magazine*, January 1993.

Saving the Everglades. *Weekly Reader*, November 4, 1994.
> Studies and attempts at conservation of the wetlands.

Stolzenburg, William. Building a Better Refuge. *Nature Conservancy*, January/February 1996.

GENERAL ARTICLES

Amazing Drugs from the Rain Forest. Drugs section, *Current Health*, March 1992.
Description of how a rain forest is used for medicines it provides.

American Biology Teacher, The. (Entire issue dedicated to environmental concerns). *The American Biology Teacher*, vol. 59, no. 6, June 1997.

Bicak, Charles J. Application of Ecological Principles in Establishing an Environmental Ethic. *The American Biology Teacher,* vol. 59, no. 4, April 1997.

Creating a Greener, Cleaner Planet Starts at Home. *Current Health,* October 1993.
Narrative about recycling at the individual level—ways kids can help.

DeFina, Anthony V. Environmental Awareness: Relating Current Issues to Biology. *The Science Teacher*, vol. 62, no. 6, September 1995.
A summary of several environmental topics.

Redmond, Tim and Marc Mowrey. Unnatural Disasters: The Ten Worst Environmental Ideas in U.S. History. Earth section, *Omni,* October 1993.
Projects that U.S. government agencies have studied, to be used to—"remedy nature's oversights."

Trash Math/Recycling Plastic: Number Please/Wanted: A Warning to Future Earthlings. Challenge, *Current Science*, March 26, 1993.
Statistics on trash by the year 2005; discussion about types of plastic; looking for a means of labeling toxic waste so that people hundreds of years from now will recognize.

Weiss, Rick. Scientists Try to Turn Weeds Into Wonder Drugs. Health section, *Washington Post,* May 10, 1994.

VIDEOS, FILMS, AND SOFTWARE BY LEARNING EXPERIENCE

Learning Experience 1

Hawkhill Science. *Ecosystems*. Hawkhill Associates, Inc., Madison, WI, 1994. (VHS 31 minutes)
This program begins with the telling of the history of ecology and perspectives of naturalists as well as hard science ecologists. It continues by outlining several major ecological concepts used to study ecosystems, and ends with a look at how human values play key roles in environmental choices.

Learning Experience 2

HRM Video. *Return to Earth: Life Cycle of the Forest.* Human Relations Media, Pleasantville, NY. (VHS, 16 minutes)

A graphic demonstration of the interdependence of living organisms in an ecosystem, this video focuses on various decomposers, and how these organisms break down organic debris that accumulates in a forest.

Learning Experience 4

Hawkhill Science. *Learning to Live in a World of Plenty, and Learning to Live in a World of Scarcity.* Hawkhill Associates, Inc., Madison, WI, 1990. (VHS, 21 minutes–Julian Simon; 27 minutes–Howard Odum)

Two videos with interviews, the first of economist Julian Simon describing why, economically, his perspective of the state of the world's population, natural resources, and pollution being manageable and sustainable; the second of ecologist Howard Odum explaining his views of a world that is running out of natural resources, seriously overpopulated by humans and in danger of collapse.

Logal Software, Inc. *Biology Explorer: Population Ecology.* P. O. Box 1499, East Arlington, MA 02174-0022. Phone 1-800-LOGAL.US (800-564-2587), available for Macintosh, Powermac, or Windows.

Biology Explorer: Population Ecology includes a guided tour and a manual. It is a very flexible program that can be used by one student, a group of students, or by the teacher as a discussion and assessment tool. The *Biology Explorer: Population Ecology* model simulates controlled ecosystems and the interrelationships between populations of organisms within each ecosystem. With representations of biotic characteristics on the organismal or cellular level, *Biology Explorer: Population Ecology* provides for in-depth investigation of organisms and the habitats in which they live. Students can model one or two habitats in the same simulation and four organisms. When the simulation is run, the program generates the dynamics of population growth and population interactions.

For Further Study of Learning Experience 4

HRM Video. *Paul Ehrlich's Earth Watch.* Human Relations Media, Pleasantville, NY. (VHS, 18 minutes)

World-renowned author and biologist, Dr. Ehrlich surveys three of the Earth's biggest environmental problems (overpopulation, mass species extinction, and global warming) and possible solutions to help save our environment.

Learning Experience 6

Washburn, Derek, Mike Cimino and Steve Bochco. *Silent Running.* Universal Pictures, 1971. (movie—90 minutes)

With a scenario of a future where the earth has been covered with asphalt and glass and the only refuge for plants and animals is on artificial biodomes in space, this dark science fiction story tells of the conflicts and difficulties of humans as caretakers of nature and its diversity of life.

For Further Study of Learning Experience 6

Queue. *Saving an Endangered Species*. Intellectual Software, division of Queue, Inc., Fairfield, CT, 1995, Intentional Educations. (Macintosh/IBM)

> This program presents information for six different simulations where you are given a role in an environmental field seeking either protection for a particular animal or plant species or approval for administering regulations, or to make decisions about environmental actions. You are given background on a fictitious committee about to take a vote on whether or not to take steps in your favor. You must weigh the information and make choices about how best to make your appeal, see the results of the appeal, and then you may write a report for analysis and evaluation.

GENERAL VIDEOS

Hawkhill Science. *Biologists at Work*. Hawkhill Associates, Inc., Madison, WI, 1994. (VHS, 34 minutes)

> Biologists at Work begins with a brief tour of the history of biology and some of its pioneers, then moves on to interviews with several outstanding biologists of today.

Hawkhill Science. *The Biosphere*. Hawkhill Associates, Inc., Madison, WI, 1994. (VHS, 42 minutes)

> The program begins by tracing the history of four billion years of the atmosphere and of life on planet earth, then gives a dramatic picture of various global environmental problems regarding our biosphere, the earth.

WEB SITES

The Globe Program
http://GLOBE.FSL.noaa.gov/

> Students and teachers from over 4,000 schools in 55 countries are working with research scientists to learn more about our planet.

South Florida Ecosystems: Changes through Time
http://fl-h2o.usgs.gov/fs95171.html

> With numerous photographs, this site gives an indepth description of the activities involved in the research of the South Florida ecosystem by the U.S. Geological Survey.

South Florida Ecosystem Fact Sheet
http://fl-h2o.usgs.gov/fs95134.html

> Using a detailed map of South Florida, this site gives an indepth description of the different aspects of the research the South Florida ecosystem by the U.S. Geological Survey.

Setting National and International Precedents in the Everglades
http://www.audubon.org/campaign/er/index.html

> A National Audobon site describing their role in the restoration of the Everglades, with links to other Audobon sites.

ORGANIZATIONS TO WRITE FOR INFORMATION

Council on Environmental Quality
722 Jackson Place, NW
Washington, DC 20006

Environmental Protection Agency
401 M St., SW
Washington, DC 20460

> Formulates and implements actions which lead to a compatible balance between human activities and the ability of natural systems to support and nurture life.

Food First
Institute for Food Development Policy
145 9th Street
San Francisco, CA 94103
415-864-8555

> Special interest group that takes measures to make countries self-sufficient in food production.

High Country News
P. O. Box 1090
Paonia, CO 81428

> Newsletters with information on grazing, mining, logging, dams, and wildlife.

Institute of Ecosystem Studies
Box AB (Route 44A)
Millbrook, NY 12545-0129

> Not-for-profit corporation dedicated to the creation, dissemination, and application of knowledge about ecological ecosystems, the institute produces scientific publications, curricula and handbooks for teachers and publishes a bimonthly newsletter.

National Audubon Society
700 Broadway
New York, NY 10003
212-979-3000

> Magazines, newsletters, and other information on environmental issues, wildlife and wildlife habitat protection; also has local chapters.

National Park Service
Department of the Interior
P. O. Box 37127
Washington, DC 20013

> Brochures, maps, pamphlets, and books covering the range of National Park lands and their ecosystems.

National Seed Storage Laboratory
Fort Collins, CO 80521
> Primary seed bank in the United States, this facility maintains backup samples of the 400,000 varieties kept in this country.

National Wildlife Federation
1400 16th Street, NW
Washington, DC 20036
202-797-6800
> Magazines, pamphlets, and other information on wildlife resources.

The Nature Conservancy
1815 N. Lynn St.
Arlington, VA 22209
703-841-5300
> Their mission is to preserve plants, animals and natural communities that represent the diversity of life on Earth by protecting the lands and waters they need to survive; magazine; also has local chapters.

Population Reference Bureau
1875 Connecticut Avenue, NW
Suite 520
Washington, DC 20009
202-483-1100
> Books, pamphlets, and tables with information about different human populations.

Seed Savers Exchange
Route 3, Box 239
Decorah, Iowa 52101
> Send a SASE for their catalogue of preserved, heirloom seeds (varieties).

Student Environmental Action Coalition (SEAC)
P. O. Box 1168
Chapel Hill, NC 27514-1168
919-967-4600
> Helps high school and college environmental groups become politically active.

Wild Earth
Earth First!
P. O. Box 455
Richmond, VT 05477
802-434-4077
> Magazine subscription includes articles on wildlife and how to protect biological diversity.

Worldwatch Papers
Worldwatch Institute
1776 Massachusetts Avenue, NW
Washington, DC 20036
202-452-1999

Pamphlets on population explosion, soil erosion, endangered species, threats to ground water.

Zero Population Growth
1400 16th Street, NW
Suite 320
Washington, DC 20036
202-332-2200

Population stabilization information.